The Feature-driven Method for Structural Optimization

The Feature-driven Method for Structural Optimization

Weihong Zhang

Ying Zhou

ELSEVIER

Elsevier
Radarweg 29, PO Box 211, 1000 AE Amsterdam, Netherlands
The Boulevard, Langford Lane, Kidlington, Oxford OX5 1GB, United Kingdom
50 Hampshire Street, 5th Floor, Cambridge, MA 02139, United States

Notices
Knowledge and best practice in this field are constantly changing. As new research and experience broaden our understanding, changes in research methods, professional practices, or medical treatment may become necessary.

Practitioners and researchers must always rely on their own experience and knowledge in evaluating and using any information, methods, compounds, or experiments described herein. In using such information or methods they should be mindful of their own safety and the safety of others, including parties for whom they have a professional responsibility.

To the fullest extent of the law, neither the Publisher nor the authors, contributors, or editors, assume any liability for any injury and/or damage to persons or property as a matter of products liability, negligence or otherwise, or from any use or operation of any methods, products, instructions, or ideas contained in the material herein.

Library of Congress Cataloging-in-Publication Data

A catalog record for this book is available from the Library of Congress

British Library Cataloguing-in-Publication Data
A catalogue record for this book is available from the British Library

ISBN: 978-0-12-821330-8

For information on all Elsevier publications
visit our website at https://www.elsevier.com/books-and-journals

Publisher: Matthew Deans
Acquisitions Editor: Glyn Jones
Editorial Project Manager: Naomi Robertson
Production Project Manager: Prem Kumar
  Kaliamoorthi
Cover Designer: Greg Harris

Typeset by SPi Global, India

Working together
to grow libraries in
developing countries

www.elsevier.com • www.bookaid.org

Contents

6. Feature-driven optimization method and applications

7. Feature-driven optimization for structures under design-dependent loads

Preface

In engineering structure optimization, developing the advanced design methodology is essential to achieving high-performance structures. One fundamental problem is how to specify design primitives that enable a structure to evolve in shape and topology. Basically, two kinds of available models can be considered for this purpose: the structural analysis model, typically such as the finite element (FE) model dedicated to evaluating the mechanical performance of the structure; and the geometric model, typically such as the CAD model dedicated to the description of the dimension and configuration of the structure. For example, earlier studies in shape optimization directly attributed the boundary node positions of the FE model as design variables to evolve the structure. In topology optimization, the so-called density-based method was also related to the FE model to remove ineffective elements. In both situations, the underlying issue is that the geometric features of an engineering structure cannot be preserved by an FE model because of its inherent discretization nature. To overcome this difficulty, postprocessing is generally needed to modify the optimized structure with specific features. From the engineering viewpoint, one identical optimized solution can be interpreted differently for the postprocessing with features. It is thus difficult to realize systematically the postprocessing without the deterioration of structural performance.

With this background, this book is dedicated to the development of the feature-driven structural optimization method for the enhancement of its practical applicability. Based on CAD modeling, the feature attributes of a structure are used as geometric primitives for the definition of design variables with the features preserved throughout the optimization process. Theories, methods, and applications are presented. The target audience of this book includes structure designers, including but not limited to mechanical, aerospace, automobile, and civil engineers; researchers and teachers in the fields of mechanical, civil, and architectural structural designs who want to keep up to date on the frontiers of structural optimization; and undergraduate and graduate students who are interested in structural optimization.

The first author greatly appreciates important contributions from his former graduate students in the past decade. Special thanks go to Dr. S.Y. Cai, L.Y. Zhao, Q.Q. Huang, Z. Xu, and L. Zhou, who worked closely with him at Northwestern Polytechnical University, China. This research work is supported by the

National Natural Science Foundation of China under the projects "Theories and methods of engineering feature-oriented structural topology optimization (11432011)" and "Advanced design theories and methods for achieving additively manufactured high-performance integrated structures (11620101002)."

Weihong Zhang
Ying Zhou
Xi'an, China

Chapter 1

Introduction

1.1 Overview and motivation

With the rapid developments of aeronautical and aerospace engineering, light-weight and high-performance structures are incessantly pursued to achieve the motto of "striving to save the weight by each gram" from aspects of both design and manufacturing. Structural optimization has experienced size, shape, and topology optimization over the last decades. Size and shape optimization aim at optimizing the size parameters and boundary shapes of a structure without topological change while topology optimization aims at optimizing the material distribution in a prescribed design domain to produce innovative structural configurations.

Topology optimization has achieved great success in both design methodology innovation and industrial applications since the introduction of the material microstructure and homogenization method. Various topology optimization methods have been developed. Among others, the density-based method, evolutionary structural optimization, and the level-set method are very popular today.

The density-based method describes the topology of a structure by representing the material distribution in terms of element density variables. To this end, interpolation models such as the solid isotropic microstructure with penalization (SIMP) model (Bendsøe and Sigmund, 1999) and the rational approximation of material properties (RAMP) model (Stolpe and Svanberg, 2001) were developed. The evolutionary structural optimization method (Xie and Steven, 1997; Tanskanen, 2002) is also a kind of density-based method that removes ineffective elements based on the intuitive criterion. Actually, the density-based method has been applied to varieties of problems, including multidisciplinary design problems, owing to its conceptual simplicity and easy implementation. Intermediate densities can be minimized or entirely eliminated by the Heaviside projection method (Guest et al., 2004) and the minimum length scale is ensured with the robust topology optimization formulation (Wang et al., 2011). Besides, a density/sensitivity filter is commonly used to prevent numerical instabilities such as checkerboard and mesh-dependence. Nevertheless, the jagged structural boundaries are inevitable for this kind of method.

The Feature-driven Method for Structural Optimization. https://doi.org/10.1016/B978-0-12-821330-8.00001-8

The level-set method originated from the moving boundary tracking problem (Osher and Sethian, 1988) and was introduced into the structural optimization field for free shape and topology changes (Wang et al., 2003; Allaire et al., 2004). This method belongs to a kind of boundary-based method and works directly with the boundary of a structure for topology optimization. Actually, two forms of level-set methods are available for the topology representation of a structure. The traditional level-set method is based on the discrete level-set function that is mesh-related with the use of level-set values at discrete nodes. This method is hindered in practical applications because of numerical complexities such as reinitialization, velocity extension, the limitation of the Courant-Friedrichs-Lewy condition, and the huge number of design variables. The continuous level-set function is hence constructed to overcome these difficulties. Wang and Wang (2006) and Luo et al. (2008) introduced the compactly supported radial basis functions (CS-RBFs) with design variables assigned by CS-RBFs' coefficients. The implicit functions of the closed form, such as R-function, KS function, and Ricci function, are also used by means of Boolean operations of geometric primitives. In this sense, the density-based method and the continuous level-set method are different in the number of design variables, the sensitivity analysis scheme, and the gray region control scheme. Moreover, the topological derivative method (Burger et al., 2004; Norato et al., 2007) could be incorporated into the level-set method for the automatic nucleation of new holes in the design domain.

At present, the common design practice is to build up the design process from structural performance optimization to feature enhancement. To have a clear idea, consider the design process of the A380 airfoil leading-edge ribs (Krog et al., 2002, 2004) shown in Fig. 1.1. The innovative topological configuration is first obtained by means of the density-based method. Then, the engineering features are manually identified from the topology optimization result. Detailed designs of the identified engineering features are further carried out through size and shape optimization until the mechanical performance is satisfied. Obviously, the artificial extraction of engineering features may change and even deteriorate the mechanical performance and original topology optimization result.

In fact, the above presentation indicates that both the element densities and discrete nodal level-set values inevitably produce a huge number of design variables. They belong to a kind of lower-level design variable with which it is impossible to preserve engineering features. Here, the feature-driven structural optimization method is developed to open up a new design avenue in terms of engineering features.

1.2   Feature-based modeling and design in CAD

For a mechanical structure, features have different engineering semantics at different stages of product development. They can be design features, manufacturing

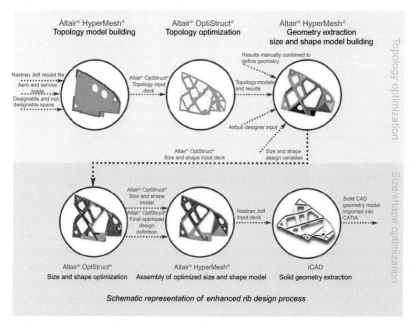

Schematic representation of enhanced rib design process

FIG. 1.1 The design process of the A380 airfoil leading-edge rib (Krog et al., 2002, 2004).

features, assembly features, etc. In the CAD community, features generally refer to shape features that are geometric objects with topological and geometric parameters in engineering structure design. Shape features could have different definitions. For example, Shah and Tadepalli (1992) stated, "A feature is a shape by which the engineers can add engineering information, attribute and knowledge for geometric reasoning." Cunningham and Dixon (1988) stated the definition as "A feature is any geometric form or entity that is used in reasoning in one or more design or manufacturing activities." Nalluri (1994) established a relatively complete shape feature library.

The majority of the concerned studies mainly focused on feature recognition and feature-based design. Feature recognition refers to the process of identifying and extracting the shape features from a geometric model based on boundary representation or constructive solid geometry. In the engineering sense, feature-based design thus takes features as basic geometric primitives in the whole process of structural modeling and design. They are instantiated, sized, and located by the designers to reason the geometry and topology of a structure (Shah and Rogers, 1993). Salomons et al. (1993) categorized the feature-based design into three stages: concept design, structure design, and parameter design.

Earlier research works of feature-based design were mainly concerned with the structural shape and size parameters (Brujic et al., 2010; Park and Dang, 2010; Zhang et al., 2010). For example, Li et al. (1999) studied the parameter design of sheet metal parts by means of the feature modeling method. Shen (2003) employed the feature recognition method to identify the beam features from a complex mechanical structure defined with polygonal meshes and optimized the identified beam features with the shape optimization method. Wang et al. (2004) carried out the feature analyses of aeroengine compressor rotor blades and developed a feature-based design system based on the UG platform. Song et al. (2007) instantiated the body features, standard parts, turbulence pole, and partition ribs in the aeroengine turbine blade. They carried out the parameter design of the turbulence pole and partition ribs.

1.3 Feature-driven structural optimization

In this book, shape features are conceptually generalized in the engineering definition. They can refer to specific portions and shaped holes of mechanical parts or embedding devices, electronic components, and modular substructures of a complicated structure system. Hence, the feature and component share the same physical meaning in terminology. The considered structure can be regarded as a multicomponent structure system. Features can be overlapped or nonoverlapped to preserve their shapes in the optimization process, depending upon their functionality and the designer's intent. In this regard, the concept of feature-based design is naturally extended to feature-driven structural optimization. The design model of a multicomponent structure is completely constructed by designable and nondesignable features to evolve the structure.

In earlier times, features were considered as rigid objects modeled by the level-set function and embedded within a structure to be topologically optimized (Qian and Ananthasuresh, 2004). This limitation was broken through by considering these features as elastic ones for load-bearing design. Meanwhile, the overlapping of the features is avoided by developing the so-called finite-circle method (Zhu and Zhang, 2006; Zhu et al., 2010; Zhang et al., 2011). Fig. 1.2 illustrates the simultaneous optimization of a multicomponent

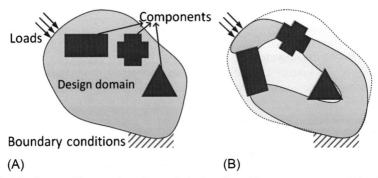

FIG. 1.2 Integrated layout and topology optimization of a multicomponent structure: (A) typical definition; and (B) optimized design (Zhu and Zhang, 2006).

layout and structural topology. The computing efficiency was further ameliorated with the superelement technique of movable components. Mei et al. (2008) proposed subtracting the simple featured holes as geometric primitives from the design domain in consideration of the machining requirement. Zhu et al. (2015) and Gao et al. (2015) treated the multicomponent structure as a double-layer system and introduced multipoint constraints to connect components with the frame structure. Chen et al. (2007) introduced the level-set function of B-splines instead of CS-RBF for the topology optimization of the frame structure. Eschenauer et al. (1994) proposed the bubble method by which the boundary shapes of the bubble features were represented as B-splines and optimized in the prescribed design domain. However, the optimization process was greatly hindered by the body-fitted remeshing of the FE model.

Alternatively, Xia et al. (2013) and Zhang et al. (2012) adopted the level-set description for the deformable solid and void components while the frame structure is described with the SIMP-based material densities. Similarly, Kang and Wang (2013) introduced featured holes with the level-set description in combination with the frame structure described by nonlocal Shepard interpolation-based material densities.

1.4   Layout of the book

This book consists of eight chapters. First, a brief introduction of the background and motivation is given. In the following chapters, the contents are focused upon level-set functions and parametric functions for structural geometric description; basic transformations and Boolean operations of level-set functions; the B-spline finite cell method for structural analysis; feature-based modeling and sensitivity analysis for structural optimization; and feature-driven optimization methods and applications. Optimization problems concerning design-dependent loads and additive manufacturing are also addressed to demonstrate the effectiveness and generality of the feature-driven structural optimization method in practical applications.

References

Allaire, G., Jouve, F., Toader, A.M., 2004. Structural optimization using sensitivity analysis and a level-set method. J. Comput. Phys. 194 (1), 363–393.

Bendsøe, M.P., Sigmund, O., 1999. Material interpolation schemes in topology optimization. Arch. Appl. Mech. 69 (9–10), 635–654.

Brujic, D., Ristic, M., Mattone, M., Maggiore, P., De Poli, G.P., 2010. CAD based shape optimization for gas turbine component design. Struct. Multidiscip. Optim. 41 (4), 647–659.

Burger, M., Hackl, B., Ring, W., 2004. Incorporating topological derivatives into level set methods. J. Comput. Phys. 194 (1), 344–362.

Chen, J., Shapiro, V., Suresh, K., Tsukanov, I., 2007. Shape optimization with topological changes and parametric control. Int. J. Numer. Methods Eng. 71 (3), 313–346.

Cunningham, J.J., Dixon, J.R., 1988. Designing with features: the origin of features. In: Proceedings of the 1988 ASME International Computers in Engineering Conference and Exhibition, July. vol. 1. ASME, New York, pp. 237–243.

Eschenauer, H.A., Kobelev, V.V., Schumacher, A., 1994. Bubble method for topology and shape optimization of structures. Struct. Optim. 8 (1), 42–51.

Gao, H.H., Zhu, J.H., Zhang, W.H., Zhou, Y., 2015. An improved adaptive constraint aggregation for integrated layout and topology optimization. Comput. Methods Appl. Mech. Eng. 289, 387–408.

Guest, J.K., Prévost, J.H., Belytschko, T., 2004. Achieving minimum length scale in topology optimization using nodal design variables and projection functions. Int. J. Numer. Methods Eng. 61 (2), 238–254.

Kang, Z., Wang, Y., 2013. Integrated topology optimization with embedded movable holes based on combined description by material density and level sets. Comput. Methods Appl. Mech. Eng. 255, 1–13.

Krog, L., Tucker, A., Rollema, G., 2002. Application of topology, sizing and shape optimization methods to optimal design of aircraft components. In: Proc. 3rd Altair UK HyperWorks Users Conference, November.

Krog, L., Tucker, A., Kemp, M., Boyd, R., 2004. Topology optimisation of aircraft wing box ribs. In: 10th AIAA/ISSMO Multidisciplinary Analysis and Optimization Conference, August.

Li, F., Zhou, X.H., Ruan, X.Y., 1999. Feature-based design of sheet metal parts. J. Mech. Sci. Technol. 18, 342–344. (in Chinese).

Luo, Z., Wang, M.Y., Wang, S., Wei, P., 2008. A level set-based parameterization method for structural shape and topology optimization. Int. J. Numer. Methods Eng. 76 (1), 1–26.

Mei, Y., Wang, X., Cheng, G., 2008. A feature-based topological optimization for structure design. Adv. Eng. Softw. 39 (2), 71–87.

Nalluri, S.R.P.R., 1994. Form Feature Generating Model for Feature Technology (Doctoral Dissertation). Indian Institute of Science, Department of Mechanical Engineering.

Norato, J.A., Bendsøe, M.P., Haber, R.B., Tortorelli, D.A., 2007. A topological derivative method for topology optimization. Struct. Multidiscip. Optim. 33 (4–5), 375–386.

Osher, S., Sethian, J.A., 1988. Fronts propagating with curvature-dependent speed: algorithms based on Hamilton-Jacobi formulations. J. Comput. Phys. 79 (1), 12–49.

Park, H.S., Dang, X.P., 2010. Structural optimization based on CAD–CAE integration and metamodeling techniques. Comput. Aided Des. 42 (10), 889–902.

Qian, Z., Ananthasuresh, G.K., 2004. Optimal embedding of rigid objects in the topology design of structures. Mech. Based Des. Struct. Mach. 32 (2), 165–193.

Salomons, O.W., van Houten, F.J., Kals, H.J.J., 1993. Review of research in feature-based design. J. Manuf. Syst. 12 (2), 113–132.

Shah, J.J., Rogers, M.T., 1993. Assembly modeling as an extension of feature-based design. Res. Eng. Des. 5 (3–4), 218–237.

Shah, J.J., Tadepalli, R., 1992. Feature based assembly modeling. In: Proceedings of the 1992 ASME International Computers in Engineering Conference and Exposition. ASME, pp. 253–260.

Shen, J., 2003. Feature-based optimization of beam structures represented by polygonal meshes. J. Comput. Inf. Sci. Eng. 3 (3), 243–249.

Song, Y.W., Hu, B.F., Xi, P., 2007. Feature-based parameterized design of turbine blade on aeroengine. Aeronaut. Manufact. Technol. 11, 73–78. (in Chinese).

Stolpe, M., Svanberg, K., 2001. An alternative interpolation scheme for minimum compliance topology optimization. Struct. Multidiscip. Optim. 22 (2), 116–124.

Tanskanen, P., 2002. The evolutionary structural optimization method: theoretical aspects. Comput. Methods Appl. Mech. Eng. 191 (47–48), 5485–5498.

Wang, S., Wang, M.Y., 2006. Radial basis functions and level set method for structural topology optimization. Int. J. Numer. Methods Eng. 65 (12), 2060–2090.

Wang, M.Y., Wang, X., Guo, D., 2003. A level set method for structural topology optimization. Comput. Methods Appl. Mech. Eng. 192 (1–2), 227–246.

Wang, R.Q., Lin, D., Fan, J., Zhu, K., 2004. Feature-based structure design of compressor blades. Aeroengine 2, 5–9. (in Chinese).

Wang, F., Lazarov, B.S., Sigmund, O., 2011. On projection methods, convergence and robust formulations in topology optimization. Struct. Multidiscip. Optim. 43 (6), 767–784.

Xia, L., Zhu, J., Zhang, W., Breitkopf, P., 2013. An implicit model for the integrated optimization of component layout and structure topology. Comput. Methods Appl. Mech. Eng. 257, 87–102.

Xie, Y.M., Steven, G.P., 1997. Basic evolutionary structural optimization. In: Evolutionary Structural Optimization. Springer, London, pp. 12–29.

Zhang, W., Wang, D., Yang, J., 2010. A parametric mapping method for curve shape optimization on 3D panel structures. Int. J. Numer. Methods Eng. 84 (4), 485–504.

Zhang, W., Xia, L., Zhu, J., Zhang, Q., 2011. Some recent advances in the integrated layout design of multicomponent systems. J. Mech. Des. 133 (10), 104503.

Zhang, J., Zhang, W.H., Zhu, J.H., Xia, L., 2012. Integrated layout design of multi-component systems using XFEM and analytical sensitivity analysis. Comput. Methods Appl. Mech. Eng. 245, 75–89.

Zhu, J.H., Zhang, W.H., 2006. Coupled design of components layout and supporting structures using shape and topology optimization. In: Proceedings of CJK-OSM IV.

Zhu, J.H., Beckers, P., Zhang, W.H., 2010. On the multi-component layout design with inertial force. J. Comput. Appl. Math. 234 (7), 2222–2230.

Zhu, J.H., Gao, H.H., Zhang, W.H., Zhou, Y., 2015. A multi-point constraints based integrated layout and topology optimization design of multi-component systems. Struct. Multidiscip. Optim. 51 (2), 397–407.

Chapter 2

Level-set functions and parametric functions

2.1 Definitions of level-set function and parametric function

2.1.1 Basic notions and geometric interpretations

Analytic geometry is a branch of geometry that studies geometric objects with algebraic methods. In analytic geometry, three kinds of functions are often used for the description of geometric objects: explicit function, parametric function, and implicit function.

- Explicit function

For an explicit function, one coordinate is explicitly defined as a function of other coordinates. To be specific, the explicit functions of planar curves and spatial surfaces can be expressed as $y = f(x)$ and $z = f(x, y)$, respectively. Two examples are given below to show the definition of explicit functions.

Example 2.1 Explicit functions of a parabola and a paraboloid
The explicit function of a parabola is expressed as

$$y = ax^2 + bx + c \qquad (2.1)$$

where a, b, and c are constants determining the location and shape of the parabola. Fig. 2.1 shows the above explicit function when $a = 1$, $b = 0$, and $c = 0$.

Similarly, the explicit function of a paraboloid with its vertex located at the origin is expressed as

$$z = \frac{x^2}{a^2} + \frac{y^2}{b^2} \qquad (2.2)$$

where a and b are constants determining the shape of the paraboloid. Eq. (2.2) with $a = b = 1$ is geometrically illustrated in Fig. 2.2.

In Figs. 2.1 and 2.2, it can be observed that the value of the dependent variable is unique for given values of independent variables. In other words, the explicit functions related to Eqs. (2.1), (2.2) are single-valued. However, most

The Feature-driven Method for Structural Optimization. https://doi.org/10.1016/B978-0-12-821330-8.00002-X

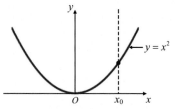

FIG. 2.1 Explicit function of a parabola ($a=1$, $b=0$, $c=0$).

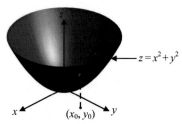

FIG. 2.2 Explicit function of a paraboloid ($a=1$, $b=1$).

geometric objects are more complex and cannot be described with single-valued functions.

Example 2.2 Explicit functions of a circle and a sphere
The explicit functions of a planar circle and a sphere can be defined with multivalued functions expressed as

$$y = \pm\sqrt{1 - x^2} \tag{2.3}$$

and

$$z = \pm\sqrt{1 - x^2 - y^2} \tag{2.4}$$

Figs. 2.3 and 2.4 show the explicit functions of the circle and sphere, respectively. It can be seen that these two explicit functions are two-valued.

However, it is often hard to obtain the closed forms of most multivalued explicit functions. As shown in Fig. 2.5, the curves of the double-back, closed, and parallel to the coordinate axis are three-valued, two-valued, and infinite-valued at the given independent variable x_0, respectively. This greatly limits the wide applications of explicit function in boundary representations for structural design.

- Parametric function

For a parametric function, each coordinate of a point is defined as a function of one or more parameters. In kinematics, parametric curves and surfaces are

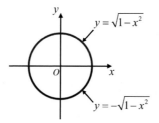

FIG. 2.3   Multivalued explicit function of a circle.

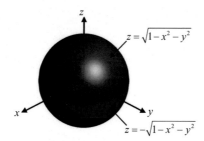

FIG. 2.4   Multivalued explicit function of a sphere.

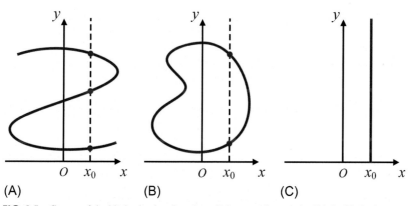

(A)   (B)   (C)

FIG. 2.5   Curves of double-back, closed, and parallel to coordinate axis: (A) double-back curve; (B) closed curve; and (C) curve parallel to y-axis.

regarded as the trajectory of a point along with the variation of parameters. The parametric function of a planar curve is generally stated as

$$P(t) = [x \ y]^T = [x(t) \ y(t)]^T \tag{2.5}$$

in which parameter t is usually related to time in kinematics. As shown in Fig. 2.6, for each parameter $t \in [t_1, t_2]$, a point $[x(t), y(t)]^T$ is determined on the curve.

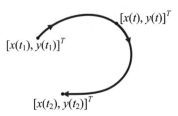

FIG. 2.6 Parametric curve.

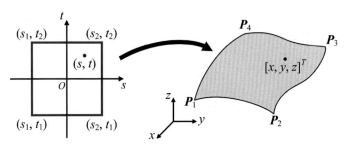

FIG. 2.7 Parametric surface.

Similarly, a spatial surface can be described with the parametric function as

$$P(s,t) = [x \ y \ z]^T = [x(s,t) \ y(s,t) \ z(s,t)]^T \tag{2.6}$$

in which s and t are two independent parameters. As shown in Fig. 2.7, a pair of parameters $(s, t) \in [s_1, s_2] \times [t_1, t_2]$ in the parametric space Ost determines a point $[x, y, z]^T$ on the surface in the Euclidean space $Oxyz$.

Notice that explicit function is a special case of parametric function where the coordinates are chosen as parameters. The explicit functions for planar curves and spatial surfaces can be rewritten in the form of parametric functions

$$P(x) = [x \ y]^T = [x \ y(x)]^T \tag{2.7}$$

and

$$P(x,y) = [x \ y \ z]^T = [x \ y \ z(x,y)]^T \tag{2.8}$$

For this reason, we mainly focus on the parametric function and implicit function in the following sections. Here, two examples are given to illuminate the parametric function.

Example 2.3 Parametric function of a circle
The parametric function of a given curve or surface is not unique depending upon how the parameterization is done. For example, the parametric function of a unit circle can be stated either trigonometrically or rationally.

- Trigonometric function

$$P(\theta) = [x(\theta) \ y(\theta)]^T = [r\cos\theta \ r\sin\theta]^T \quad \theta \in [0, 2\pi] \tag{2.9}$$

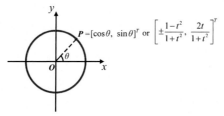

FIG. 2.8 Parametric functions of a unit circle.

where $r = 1$ denotes the radius of the unit circle. Here, parameter θ is related to the angle between vectors $\overrightarrow{OP}$ and $\overrightarrow{Ox}$, as shown in Fig. 2.8.

- Rational function

$$P(t) = [x(t) \; y(t)]^T = \left[\pm \frac{1-t^2}{1+t^2} \quad \frac{2t}{1+t^2} \right]^T \quad t \in [-1, 1] \tag{2.10}$$

where parameter t does not have the physical meaning.

Example 2.4 Parametric function of a torus
The parametric function of a torus is stated as

$$P(\varphi, \theta) = \begin{bmatrix} x(\varphi, \theta) \\ y(\varphi, \theta) \\ z(\varphi, \theta) \end{bmatrix} = \begin{bmatrix} (R + r\cos\varphi)\cos\theta \\ (R + r\cos\varphi)\sin\theta \\ r\sin\varphi \end{bmatrix} \tag{2.11}$$

where φ and θ are two parameters that vary between 0 and 2π. Fig. 2.9 shows the torus with $R = 2$ and $r = 1/2$.

With the parametric function, it is easy to obtain the points on the boundary of a geometric object. In modern CAD systems, parametric curves and surfaces such as Bezier, B-spline, and NURBS curves/surfaces are widely used for geometric modeling.

- Implicit function

Alternatively, a geometric object can be described by means of the zero level-set of a higher-dimensional implicit function $\Phi(x)$. As shown in Fig. 2.10, the

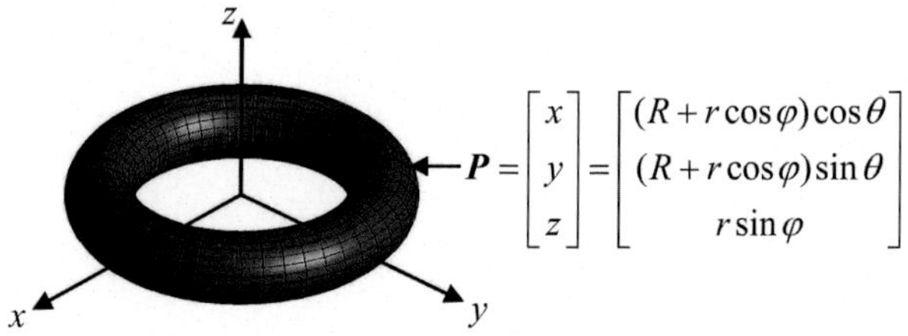

$$P = \begin{bmatrix} x \\ y \\ z \end{bmatrix} = \begin{bmatrix} (R + r\cos\varphi)\cos\theta \\ (R + r\cos\varphi)\sin\theta \\ r\sin\varphi \end{bmatrix}$$

FIG. 2.9 Parametric function of a torus ($R = 2$, $r = 1/2$).

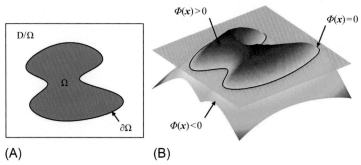

FIG. 2.10 Geometric representation of the domain Ω with a level-set function $\Phi(x)$: (A) geometric domain Ω; and (B) level-set function $\Phi(x)$.

boundary of domain Ω in design domain D can be stated as $\{x \in D \,|\, \Phi(x) = 0\}$. The implicit function in analytic geometry is also called "level-set function" in structural design. Herein, the terminology "level-set function" will be used if not specified.

Unlike the parametric function, the level-set function $\Phi(x)$ does not include the analytic information of the coordinates of the boundary points $x \in \partial\Omega$. However, the relative position of an arbitrary point $x \in D$ can simply be identified according to the sign of $\Phi(x)$.

$$\begin{cases} \Phi(x) > 0 & \forall x \in \Omega \\ \Phi(x) = 0 & \forall x \in \partial\Omega \\ \Phi(x) < 0 & \forall x \in D \backslash \bar{\Omega} \end{cases} \tag{2.12}$$

Notice that for a given geometric object, the zero-contour represented by different level-set functions could be the same. In other words, the level-set function of a given geometric object is not unique. Two examples are illustrated below to highlight the properties of level-set functions.

Example 2.5 Level-set functions of a circle
Consider four level-set functions of an identical circle centered at the origin: quadratic form $\Phi_1(x)$, signed distance form $\Phi_2(x)$ (Fougerolle et al., 2005), normalized form $\Phi_3(x)$, and logarithmic form $\Phi_4(x)$ (Fougerolle et al., 2013).

$$\begin{cases} \Phi_1(x, y) = r^2 - (x^2 + y^2) \\ \Phi_2(x, y) = r - \sqrt{x^2 + y^2} \\ \Phi_3(x, y) = 1 - \dfrac{x^2 + y^2}{r^2} \\ \Phi_4(x, y) = \ln\left(\dfrac{r^2}{x^2 + y^2}\right) \end{cases} \tag{2.13}$$

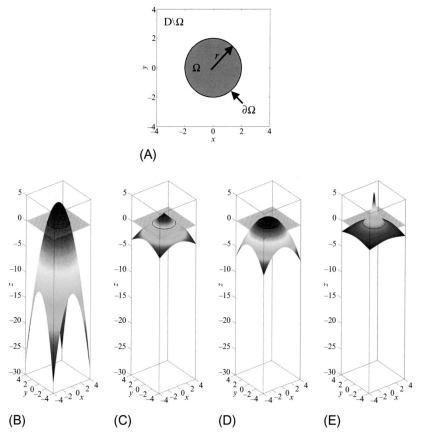

FIG. 2.11 Four level-set functions of an identical circle: (A) circular domain Ω in the design domain $D = [-4, 4] \times [-4, 4]$; (B) quadratic form $\Phi_1(x, y)$; (C) signed distance form $\Phi_2(x, y)$; (D) normalized form $\Phi_3(x, y)$; and (E) logarithmic form $\Phi_4(x, y)$.

where r denotes the radius of the circle. As shown in Fig. 2.11, the above level-set functions are different in higher-dimensional space. However, their zero level-sets represent the same circle.

Example 2.6 Level-set function of cassini oval

The level-set function of the cassini oval is stated as

$$\Phi(x, y) = \left(x^2 + y^2\right)^2 - 2b^2\left(x^2 - y^2\right) - \left(a^4 - b^4\right) \tag{2.14}$$

where a and b are parameters controlling its shape and topology. The higher-dimensional level-set function $\Phi(x, y)$ and its zero level-set evolve along with the variation of a and b, as shown in Fig. 2.12. It can be seen that the level-set function is capable of handling topology change freely, which is one of the most attractive properties of the level-set function.

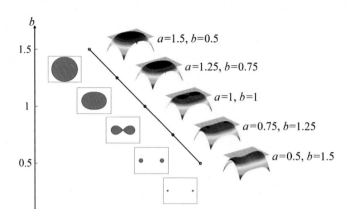

FIG. 2.12 Evolution of the cassini oval and its higher-dimensional level-set function.

2.1.2   Gradient, curvature, and convexity

In this section, basic quantities will be introduced to characterize both parametric and level-set functions.

- Parametric function

In mathematics, the tangent vector of a planar curve is the vector tangent to this curve. For the parametric curve related to Eq. (2.5), its unit tangent vector t can be expressed as

$$t = \frac{\boldsymbol{P}'(t)}{\|\boldsymbol{P}'(t)\|} = \frac{1}{\sqrt{x'(t)^2 + y'(t)^2}} [x'(t)\ \ y'(t)]^T \tag{2.15}$$

The normal vector measures the variation of the tangent vector along the curve. It is perpendicular to the tangent vector. The unit normal vector n of the parametric curve is computed by

$$n = \frac{1}{\sqrt{x'(t)^2 + y'(t)^2}} [y'(t)\ \ -x'(t)]^T \tag{2.16}$$

Besides, the curvature of the parametric curve is calculated by

$$\kappa = \frac{\|\boldsymbol{P}'(t) \times \boldsymbol{P}''(t)\|}{\|\boldsymbol{P}''(t)\|^3} = \frac{x'(t)y''(t) - x''(t)y'(t)}{\left[x''(t)^2 + y''(t)^2\right]^{\frac{3}{2}}} \tag{2.17}$$

For the parametric surface defined by Eq. (2.6), its tangent surface is determined by two tangent vectors, t_1 and t_2, which are expressed as

$$
\begin{cases}
t_1 = \left[\dfrac{\partial x(s,t)}{\partial s} \quad \dfrac{\partial y(s,t)}{\partial s} \quad \dfrac{\partial z(s,t)}{\partial s} \right]^T \\[2ex]
t_2 = \left[\dfrac{\partial x(s,t)}{\partial t} \quad \dfrac{\partial y(s,t)}{\partial t} \quad \dfrac{\partial z(s,t)}{\partial t} \right]^T
\end{cases}
\tag{2.18}
$$

The normal vector is perpendicular to the tangent surface. The unit normal vector of the parametric surface reads

$$
n = \frac{t_1 \times t_2}{\| t_1 \times t_2 \|}
\tag{2.19}
$$

There exist an infinite number of normal surfaces passing through a point on the parametric surface. The curvature of the intersection line between the parametric and normal surfaces is the normal curvature of this point. In this sense, a point has an infinite number of curvatures. Among others, the maximum and minimum curvatures are the principal curvatures denoted by κ_1 and κ_2. The product of κ_1 and κ_2 is called the Gauss curvature, measuring the bent degree of the parametric surface.

- Implicit function

The gradient of a multivariate function is a vector whose components are the first-order derivatives of the function with respect to the involved variables. The gradient of the level-set function $\Phi(x){:}R^n \to R$ is expressed as

$$
\nabla \Phi = \left[\frac{\partial \Phi}{\partial x_1}, \frac{\partial \Phi}{\partial x_2}, \dots, \frac{\partial \Phi}{\partial x_n} \right]^T
\tag{2.20}
$$

The unit normal vector n is then computed by

$$
n = \frac{\nabla \Phi}{\| \nabla \Phi \|}
\tag{2.21}
$$

The normal vector points to the normal direction of the boundary described by the level-set function. Notice that the direction of the normal vector is not unique. At a given point on a curve or a surface, two oppositely directed normal vectors exist. Inward normal and outward normal are distinguished according to whether the normal vector points to the inside or outside of a closed curve or surface.

The curvature of the implicit curve or surface is calculated by

$$
\kappa = -\operatorname{div}(n) = -\operatorname{div}\left(\frac{\nabla \Phi}{\| \nabla \Phi \|} \right)
\tag{2.22}
$$

The curvature is generally used to judge the convexity of a closed region. The closed region is convex only if the sign of curvature κ remains unchanged along the boundary.

Example 2.7 Gradient, curvature, and convexity of a circle
A circle is used to illustrate the calculations of gradient and curvature.

- Parametric function

First, consider the parametric function defined by Eq. (2.9). Its unit tangent vector is computed as

$$t = \frac{1}{\sqrt{x'(\theta)^2 + y'(\theta)^2}}[x'(\theta) \quad y'(\theta)]^T = [-\sin\theta \quad \cos\theta]^T \qquad (2.23)$$

The unit normal vector is calculated by

$$n = \frac{1}{\sqrt{x'(\theta)^2 + y'(\theta)^2}}[y'(\theta) \quad -x'(\theta)]^T = [\cos\theta \quad \sin\theta]^T \qquad (2.24)$$

The curvature of the parametric function is computed by

$$\kappa = \frac{x'(\theta)y''(\theta) - x''(\theta)y'(\theta)}{\left[x''(\theta)^2 + y''(\theta)^2\right]^{\frac{3}{2}}} = \frac{r^2}{(r^2)^{\frac{3}{2}}} = \frac{1}{r} \qquad (2.25)$$

- Level-set function

Second, two different level-set functions, Φ_1 and Φ_2, are considered. For level-set function $\Phi_1(x, y)$ in Eq. (2.13), the gradient is computed as

$$\nabla\Phi_1 = \left[\frac{\partial\Phi_1}{\partial x} \quad \frac{\partial\Phi_1}{\partial y}\right]^T = [-2x \quad -2y]^T \qquad (2.26)$$

The unit normal vector is calculated as

$$n_1 = \frac{\nabla\Phi_1}{\|\nabla\Phi_1\|} = \left[-\frac{x}{\sqrt{x^2 + y^2}} \quad -\frac{y}{\sqrt{x^2 + y^2}}\right]^T \qquad (2.27)$$

The curvature of the circle corresponds to

$$\kappa_1 = -\text{div}\left(\frac{\nabla\Phi_1}{\|\nabla\Phi_1\|}\right) = -\frac{\left(\frac{\partial\Phi_1}{\partial x}\right)^2\frac{\partial^2\Phi_1}{\partial y^2} - 2\frac{\partial\Phi_1}{\partial x}\frac{\partial\Phi_1}{\partial y}\frac{\partial^2\Phi_1}{\partial x\partial y} + \frac{\partial^2\Phi_1}{\partial x^2}\left(\frac{\partial\Phi_1}{\partial y}\right)^2}{\|\nabla\Phi_1\|^3}$$

$$= \frac{1}{\sqrt{x^2 + y^2}} \qquad (2.28)$$

Similar results can be obtained for the level-set function $\Phi_2(x, y)$ in Eq. (2.13).

$$\nabla \Phi_2 = \left[\frac{\partial \Phi_2}{\partial x} \quad \frac{\partial \Phi_2}{\partial y} \right]^T = \left[-\frac{x}{\sqrt{x^2+y^2}} \quad -\frac{y}{\sqrt{x^2+y^2}} \right]^T \tag{2.29}$$

$$n_2 = \frac{\nabla \Phi_2}{\|\nabla \Phi_2\|} = \left[-\frac{x}{\sqrt{x^2+y^2}} \quad -\frac{y}{\sqrt{x^2+y^2}} \right]^T \tag{2.30}$$

$$\kappa_2 = -\mathrm{div}\left(\frac{\nabla \Phi_2}{\|\nabla \Phi_2\|} \right) = \frac{1}{\sqrt{x^2+y^2}} \tag{2.31}$$

Notice that the unit normal vector n and the curvature κ calculated with different level-set functions are unchanged. Besides, the sign of curvature κ remains unchanged along the circle's boundary, which indicates that the circle is convex.

2.2 Heaviside function, Dirac delta function, and regularized forms

2.2.1 Basic notions

The Heaviside function, also called the Heaviside step function, is a discontinuous function. As illustrated in Fig. 2.13, it values zero for negative input and one for nonnegative input.

$$H(t) = \begin{cases} 1, & t \geq 0 \\ 0, & t < 0 \end{cases} \tag{2.32}$$

The Heaviside function is often used in combination with the level-set function of a geometric object. By substituting the level-set function Φ into Eq. (2.32), a mapping $H : \Phi \to \{0,1\}$ is obtained to compress the level-set function into a characteristic function $H(\Phi)$, as shown in Fig. 2.14. In structural optimization, the characteristic function $H(\Phi)$ is usually used to indicate the material distribution for the description of structural topology. To be specific, the structural domain Ω is occupied with solid material for $\Phi(x) \geq 0$ while the domain where $\Phi(x) < 0$ is assigned with void material.

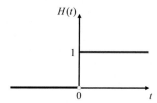

FIG. 2.13 Heaviside function.

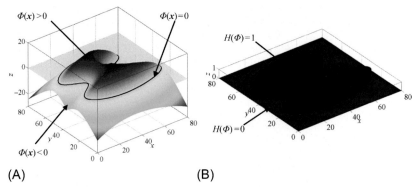

FIG. 2.14 Heaviside mapping from the level-set function Φ to the characteristic function $H(\Phi)$: (A) level-set function $\Phi(x)$; and (B) characteristic function $H(\Phi)$.

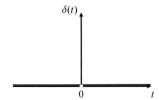

FIG. 2.15 Dirac delta function.

As illustrated in Fig. 2.15, the derivative of the Heaviside function is the Dirac delta function, which is usually denoted as the δ-function. It values zero everywhere except at the origin point $t = 0$.

$$\delta(t) = \begin{cases} +\infty, & t = 0 \\ 0, & t \neq 0 \end{cases} \tag{2.33}$$

2.2.2 Regularized Heaviside function and Dirac delta function

In many applications, the discontinuity of the Heaviside function is unexpected and usually brings numerical issues when calculating the derivatives. Therefore, variants of regularized forms $\hat{H}(t)$ have been proposed in many of the literature works to regularize the Heaviside function in Eq. (2.32). The regularization is achieved by a smooth transition from 0 to 1 around the discontinuous region, as shown in Fig. 2.16A. In this sense, the material distribution describing the structural topology has a smooth transition around the structural boundary due to the regularized characteristic function $\hat{H}(\Phi)$. To be specific, the domain is occupied with solid material for $\Phi(x) \geq \Delta$; the domain with $\Phi(x) < -\Delta$ is

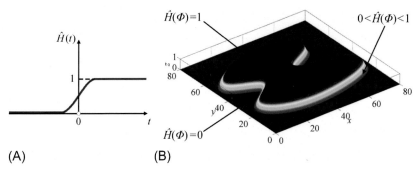

FIG. 2.16   Regularization of the Heaviside function and the corresponding smoothed characteristic function: (A) regularized Heaviside function $\hat{H}(t)$; and (B) smoothed characteristic function $\hat{H}(\Phi)$.

assigned with void material; and the intermediate material exists within the narrow-band domain for $-\Delta \leq \Phi(x) < \Delta$.

Table 2.1 is a list of typical regularizations of the Heaviside function and the Dirac delta function. In all expressions, Δ affects the width of the transition domain.

Example 2.8 Effects of Δ on the regularized Heaviside function and the Dirac delta function

The regularized form 2 in Table 2.1 is used to study the effects of parameter Δ on the smooth transition. By changing the value of Δ, the corresponding regularized Heaviside function and Dirac delta function are illustrated in Figs. 2.17 and 2.18, respectively. It is shown that the transition domain is enlarged and the peak value of the regularized Dirac delta function decreases as Δ increases.

2.3 Typical level-set functions

2.3.1 Signed distance function and first-order approximation

The signed distance function (SDF) is a typical form of the level-set function that is defined as

$$\Phi_{SDF}(x) = \begin{cases} d(x), & x \in \Omega \\ 0, & x \in \partial\Omega \\ -d(x), & x \in D/\bar{\Omega} \end{cases} \tag{2.34}$$

in which $d(x)$ refers to the minimum distance of point x to boundary $\partial\Omega$.

$$d(x) = \min_{x_p \in \partial\Omega} \|x - x_p\|, \quad x \in D \tag{2.35}$$

The signed distance function has the property of the unit gradient module with $\|\nabla \Phi_{SDF}\| = 1$. To give an example, the level-set function $\Phi_2(x, y)$ in Eq. (2.13) is the signed distance function of the circle. Geometrically, it means

TABLE 2.1 Regularized forms of Heaviside function and δ-function.

No.	Regularized Heaviside function	Regularized δ-function	References
1	$\hat{H}(t) = \begin{cases} 1, & t \geq \Delta \\ \frac{1}{2} + \frac{t}{2\Delta}, & -\Delta \leq t < \Delta \\ 0, & t < -\Delta \end{cases}$	$\hat{\delta}(t) = \begin{cases} \frac{1}{2\Delta}, & -\Delta \leq t < \Delta \\ 0, & \text{otherwise} \end{cases}$	Kumar et al. (2008)
2	$\hat{H}(t) = \begin{cases} 1, & t \geq \Delta \\ \frac{3}{4}\left(\frac{t}{\Delta} - \frac{t^3}{3\Delta^3}\right) + \frac{1}{2}, & -\Delta \leq t < \Delta \\ 0, & t < -\Delta \end{cases}$	$\hat{\delta}(t) = \begin{cases} \frac{3}{4}\left(\frac{1}{\Delta} - \frac{t^2}{\Delta^2}\right), & -\Delta \leq t < \Delta \\ 0, & \text{otherwise} \end{cases}$	Wang et al. (2003) and Wang and Wang (2004)
3	$\hat{H}(t) = \begin{cases} 1, & t \geq \Delta \\ \frac{1}{2} + \frac{15t}{16\Delta} - \frac{5t^3}{8\Delta^3} + \frac{3t^5}{16\Delta^5}, & -\Delta \leq t < \Delta \\ 0, & t < -\Delta \end{cases}$	$\hat{\delta}(t) = \begin{cases} \frac{15}{16}\left(\frac{1}{\Delta} - \frac{2t^2}{\Delta^3} + \frac{t^4}{\Delta^5}\right), & -\Delta \leq t < \Delta \\ 0, & \text{otherwise} \end{cases}$	Kawamoto et al. (2011)
4	$\hat{H}(t) = \begin{cases} 1, & t \geq \Delta \\ \frac{1}{2}\left(1 + \sin\frac{\pi t}{2\Delta}\right), & -\Delta \leq t < \Delta \\ 0, & t < -\Delta \end{cases}$	$\hat{\delta}(t) = \begin{cases} \frac{\pi}{4\Delta}\cos\frac{\pi t}{2\Delta}, & -\Delta \leq t < \Delta \\ 0, & \text{otherwise} \end{cases}$	Zhang et al. (2012)
5	$\hat{H}(t) = \begin{cases} 1, & t \geq \Delta \\ \frac{1}{2}\left(1 + \frac{t}{\Delta} + \frac{1}{\pi}\sin\frac{\pi t}{\Delta}\right), & -\Delta \leq t < \Delta \\ 0, & t < -\Delta \end{cases}$	$\hat{\delta}(t) = \begin{cases} \frac{1}{2\Delta}\left(1 + \cos\frac{\pi t}{\Delta}\right), & -\Delta \leq t < \Delta \\ 0, & \text{otherwise} \end{cases}$	Zhao et al. (1996)
6	$\hat{H}(t) = \frac{1}{2} + \frac{1}{\pi}\arctan\left(\frac{t}{\Delta}\right)$	$\hat{\delta}(t) = \frac{\Delta}{\pi(\Delta^2 + t^2)}$	Xia et al. (2013)
7	$\hat{H}(t) = \frac{1}{2} + \frac{1}{2}\tanh\left(\frac{t}{\Delta}\right) = \frac{1}{1 + e^{-2t/\Delta}}$	$\hat{\delta}(t) = \frac{1}{2\Delta}\left(1 - \tanh^2\left(\frac{t}{\Delta}\right)\right) = \frac{2}{\Delta}\frac{e^{-2t/\Delta}}{\left(1 + e^{-2t/\Delta}\right)^2}$	Allaire et al. (2016)
8	$\hat{H}(t) = \frac{1}{2} + \frac{1}{2}\text{erf}\left(\frac{t}{\Delta}\right)$	$\hat{\delta}(t) = \frac{1}{\sqrt{\pi}\Delta}e^{-\frac{t^2}{\Delta}}$	Andrews (1998)

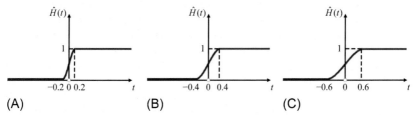

FIG. 2.17 Effects of Δ on the regularized form 2 of the Heaviside function in Table 2.1: (A) Δ = 0.2; (B) Δ = 0.4; and (C) Δ = 0.6.

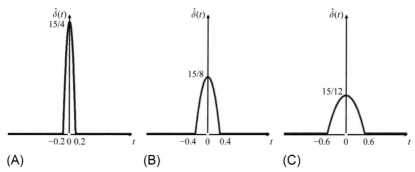

FIG. 2.18 Effects of Δ on the regularized form 2 of the δ-function in Table 2.1: (A) Δ = 0.2; (B) Δ = 0.4; and (C) Δ = 0.6.

that the Δ-contour of the signed distance function is the offset of its zero-contour along the normal direction and the offset distance equals Δ. The signed distance function plays an important role in many applications such as surface reconstruction, scientific visualization, structural optimization, etc.

The material transition depends upon the nature of the level-set function $\Phi(x)$. Different forms of $\Phi(x)$ will bring out distinct material transition distributions. Theoretically, the narrow-band transition domain is quite broad at the place where the high-dimensional level-set function is flat and it becomes quite thin at the place where the high-dimensional level-set function is sharp. This implies that the nonequidistant distribution of the material transition around the zero-contour of the structure will bring out the fuzziness and disagreement between the geometric model and the material model of a structure. If the signed distance function is introduced, the narrow-band transition domain occupied with intermediate material will have a uniform width that equals 2Δ everywhere. Therefore, the level-set function should be properly selected.

In fact, the material transition width should be distance-dependent instead of function value-dependent. Otherwise, the nonequidistant distribution of the material transition takes place around the structural boundaries. For this reason, the first-order approximation is introduced to normalize the level-set function into a quasiequidistant iso-contour function.

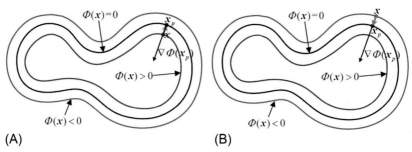

FIG. 2.19 First-order approximation of the level-set function $\Phi(x)$ into signed distance function: (A) x is the offset of x_p in the direction of inward normal; and (B) x is the offset of x_p in the direction of outward normal.

Consider a level-set function of the domain in Fig. 2.19A. Suppose point x is the offset of a point x_p on the structural boundary in the direction of inward normal and the offset distance is $\|x_p - x\|$. The following relation holds based on the Taylor expansion (Taubin, 1994)

$$\Phi(x_p) = \Phi(x) + \nabla^T \Phi(x)(x_p - x) + o(\|x_p - x\|) \tag{2.36}$$

As shown in Fig. 2.19A, the vectors $\nabla \Phi(x)$ and $x_p - x$ are in opposite directions, that is, their relative angle equals π; the second term in Eq. (2.36) can thus be written as

$$\nabla^T \Phi(x)(x_p - x) = \|\nabla\Phi(x)\| \cdot \|x_p - x\| \cdot \cos\pi = -\|\nabla\Phi(x)\| \cdot \|x_p - x\| \tag{2.37}$$

Substitute Eq. (2.37) into (2.36) and neglect the second-order small quantity. The following relation can be deduced due to the equality $\Phi(x_p) = 0$

$$0 = \Phi(x_p) \approx \Phi(x) - \|\nabla\Phi(x)\| \cdot \|x_p - x\| \tag{2.38}$$

Thus, we can obtain that

$$\|x_p - x\| = \frac{\Phi(x)}{\|\nabla\Phi(x)\|} \tag{2.39}$$

Similarly, suppose x is the offset of x_p in the direction of outward normal as shown in Fig. 2.19B. The vectors $\nabla \Phi(x)$ and $x_p - x$ are in the same direction so that the following approximation holds

$$0 = \Phi(x_p) \approx \Phi(x) + \|\nabla\Phi(x)\| \cdot \|x_p - x\| \tag{2.40}$$

In this case, the signed distance function is approximately written as

$$-\|x_p - x\| = \frac{\Phi(x)}{\|\nabla\Phi(x)\|} \tag{2.41}$$

In summary, the first-order approximation of $\Phi(x)$ into signed distance function reads

$$\overline{\Phi}(x) = \frac{\Phi(x)}{\|\nabla\Phi(x)\|} \tag{2.42}$$

One-dimensional (1D) and two-dimensional (2D) examples are given below to show the quasiequidistant effects of the first-order approximations in comparison with the original level-set functions.

Example 2.9 First-order approximation to describe interval $[-4, 4]$ with signed distance function

Consider a closed interval $[-4, 4]$ whose signed distance function $\Phi_{SDF}(x)$ and quadratic level-set function $\Phi(x)$ are

$$\Phi_{SDF}(x) = 4 - |x| \qquad (2.43)$$

and

$$\Phi(x) = 1 - \frac{x^2}{16} \qquad (2.44)$$

According to Eq. (2.42), the first-order approximation of $\Phi(x)$ into signed distance function is computed as

$$\overline{\Phi}(x) = \frac{\Phi(x)}{\|\nabla\Phi(x)\|} = \frac{16 - x^2}{2|x|} \qquad (2.45)$$

Although these functions have the same root ($x = \pm 4$) representing the boundary of the interval, they behave differently when compressed by the regularized Heaviside function because of the differences of iso-contour distributions. To clarify the idea, Fig. 2.20 highlights three smoothed characteristic functions $\hat{H}(\Phi_{SDF})$, $\hat{H}(\Phi)$, and $\hat{H}(\overline{\Phi})$ within the narrow-band $[-\Delta, \Delta]$ at $\Delta = 0.5$. The quadratic level-set function $\Phi(x)$ is flatter than $\Phi_{SDF}(x)$ around the interval boundary $x = \pm 4$. That is, the narrow-band interval $[-\Delta, \Delta]$ of $\Phi(x)$ results in a broader interval of x than that of $\Phi_{SDF}(x)$ while the first-order approximation increases the slope value around the interval boundary $x = \pm 4$ so that the corresponding interval width of x related to $[-\Delta, \Delta]$ reduces immensely to favor the compression effect of the regularized Heaviside function.

Example 2.10 First-order approximation to describe a circle with signed distance function

Consider a circle centered at the origin with radius r. The level-set function $\Phi_2(x, y)$ in Eq. (2.13) corresponds to its signed distance function $\Phi_{SDF}(x, y)$. Fig. 2.21A illustrates that $\Phi_{SDF}(x, y)$ is fully normalized and equidistant with a unit gradient module in all directions. In contrast, Fig. 2.21B indicates that $\Phi_3(x, y)$ in Eq. (2.13) is flatter inside the boundary than outside the boundary. Fig. 2.21C depicts its first-order approximation $\overline{\Phi}(x, y)$. Correspondingly, Fig. 2.22 shows the iso-contours of the three functions between $[-0.8, 0.8]$ and their compressed functions $\hat{H}(\Phi_{SDF})$, $\hat{H}(\Phi_3)$, and $\hat{H}(\overline{\Phi})$ along the radial direction. As expected, Fig. 2.23B conforms that the nonequidistant property of the iso-contours of $\Phi_3(x, y)$ enlarges the narrow-band along the radial direction so that a strong distribution of intermediate materials around $\Phi_3(x, y) = 0$ occurs. Fortunately, Fig. 2.23C shows that the first-order approximation $\overline{\Phi}(x, y)$ greatly improves the distribution of intermediate materials.

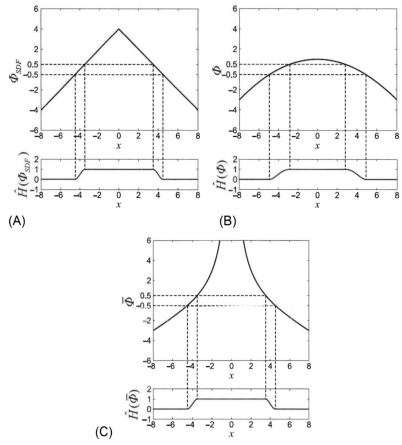

FIG. 2.20 Level-set functions of interval $[-4, 4]$ and their effects upon the smoothed characteristic function within the narrow-band interval $\Phi \in [-\Delta, \Delta]$: (A) signed distance function $\Phi_{SDF}(x)$; (B) quadratic level-set function $\Phi(x)$; and (C) first-order approximation $\overline{\Phi}(x)$.

Example 2.11 First-order approximation to describe a superellipse with signed distance function

The level-set function of a superellipse can be expressed as

$$\Phi(x, y) = 1 - \sqrt[n]{\left|\frac{\cos\theta(x - x_0) + \sin\theta(y - y_0)}{a}\right|^n + \left|\frac{-\sin\theta(x - x_0) + \cos\theta(y - y_0)}{b}\right|^n}$$

$$(2.46)$$

where (x_0, y_0) denotes the reference center of the superellipse; θ refers to the inclined angle of the superellipse; and a and b are the semilength and semiwidth,

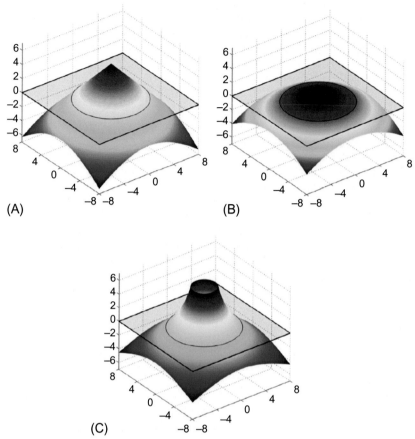

FIG. 2.21 Higher-dimensional level-set functions of the circle: (A) signed distance function Φ_{SDF} (x, y); (B) normalized quadratic level-set function $\Phi_3(x, y)$ in Eq. (2.13); and (C) first-order approximation $\overline{\Phi}(x, y)$.

respectively. n controls the degree of rounded corners of the superellipse. The larger n is, the sharper the corners are.

As the closed-form of its signed distance function is not available, Eq. (2.42) is used to derive the corresponding first-order approximation. Fig. 2.24 gives the higher-dimensional representations of both. Likewise, Fig. 2.25 shows the iso-contours of both functions between $[-0.4, 0.4]$ and their smoothed characteristic functions $\hat{H}(\Phi)$ and $\hat{H}(\overline{\Phi})$ along the x and y directions. It is observed that the flatness of $\Phi(x, y)$ with a nonequidistant iso-contour is greatly improved by the first-order approximation with a quasiequidistant iso-contour. From Fig. 2.26, one can draw out the same conclusion that intermediate materials are reduced by the first-order approximation to a great extent.

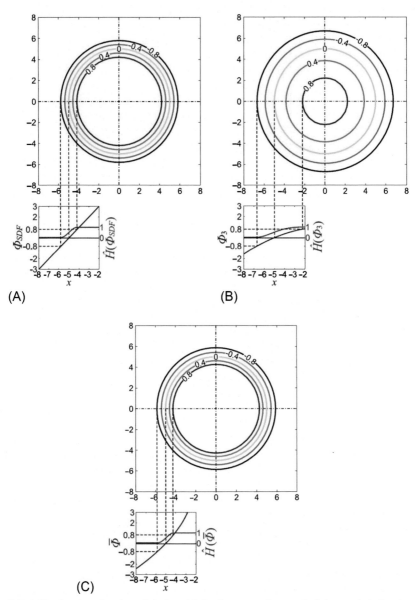

FIG. 2.22 Level-set functions of circle and their effects upon the smoothed characteristic function within the narrow-band interval $\Phi \in [-\Delta, \Delta]$: (A) signed distance function $\Phi_{SDF}(x, y)$; (B) normalized quadratic level-set function $\Phi_3(x, y)$ in Eq. (2.13); and (C) first-order approximation $\overline{\Phi}(x, y)$.

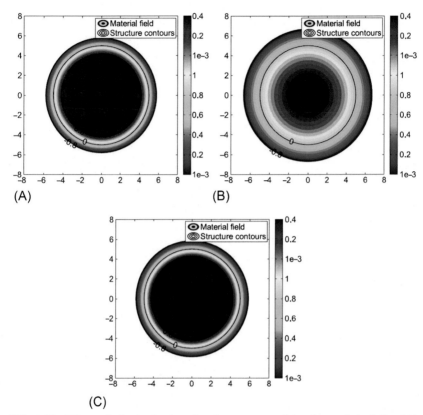

FIG. 2.23 Effects of circular level-set functions compressed by the regularized Heaviside function upon the distribution of intermediate materials: (A) signed distance function $\Phi_{SDF}(x, y)$; (B) normalized quadratic level-set function $\Phi_3(x, y)$ in Eq. (2.13); and (C) first-order approximation $\overline{\Phi}(x, y)$.

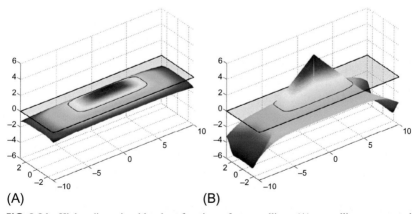

FIG. 2.24 Higher-dimensional level-set functions of a superellipse: (A) superellipse represented by the level-set function $\Phi(x, y)$; and (B) superellipse represented by the first-order approximation $\overline{\Phi}(x, y)$.

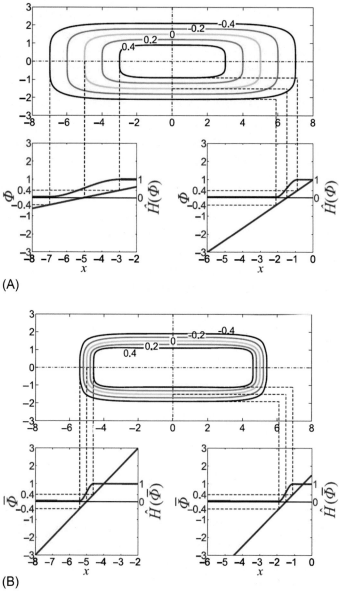

FIG. 2.25 Level-set functions of superellipse and their effects upon the smoothed characteristic function within the narrow-band interval $[-\Delta, \Delta]$: (A) superellipse represented by the level-set function $\Phi(x, y)$; and (B) superellipse represented by the first-order approximation $\overline{\Phi}(x, y)$.

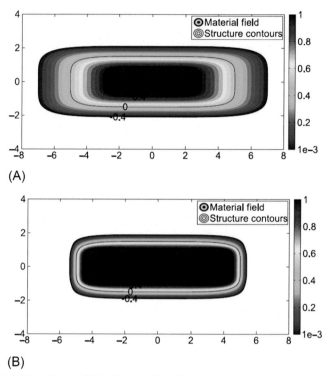

(A)

(B)

FIG. 2.26 Effects of superelliptical level-set functions compressed by the regularized Heaviside function upon the distribution of intermediate materials: (A) superellipse represented by the level-set function $\Phi(x, y)$; and (B) superellipse represented by the first-order approximation $\overline{\Phi}(x, y)$.

2.3.2   Radial basis function

The radial basis function (RBF), popular in scattered data fitting and function approximation, can naturally be used to approximate the discrete level-set function. RBFs can commonly be divided into two categories: globally supported RBFs (GS-RBFs) and compactly supported RBFs (CS-RBFs) (Buhmann, 2000; Carr et al., 2001; Ohtake et al., 2005; Skala, 2012). Table 2.2 lists some typical forms of GS-RBFs.

In Table 2.2, $[k]$ denotes the maximum integer smaller than k. N is a natural number. R^+ is the set of all positive real numbers. c is an arbitrary constant. The conditionally positive definiteness order indicates the solvability of the interpolation. r denotes the radius of support related to point (x_i, y_i) and is usually defined by (if $\boldsymbol{x} = \{x, y\}^T \in R^2$)

$$r = \frac{d_i}{d_p} = \frac{\sqrt{(x - x_i)^2 + (y - y_i)^2}}{d_p} \tag{2.47}$$

TABLE 2.2 Typical forms of GS-RBFs.

GS-RBFs	$\varphi(x)$	Conditionally positive definiteness order
Power spline (PS)	$r^{2k+1},\ k \in N$	$k+1$
Thin-plate splines (TPS)	$r^{2k}\ln(r),\ k \in N$	$k+1$
Multiquadric (MQ)	$(r^2+c^2)^k,\ k \in R^+$	$[k]$
Inverse MQ	$(r^2+c^2)^{-k},\ k \in R^+$	0
Gaussian	$e^{-\frac{r^2}{c^2}}$	0
Markoff	$e^{-c\lvert r \rvert}$	/

where d_p is the predefined support radius that should be chosen appropriately. (x_i, y_i) is the coordinate of the ith knot, that is, the center of the ith RBF's support.

The radial basis functions listed in Table 2.2 are illustrated in Fig. 2.27.

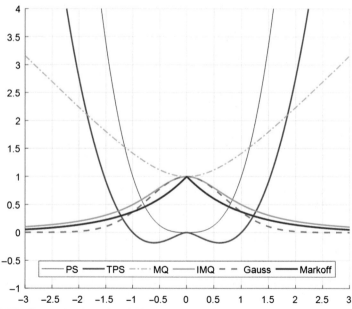

FIG. 2.27 Geometric schematics of GS-RBFs listed in Table 2.2.

GS-RBFs usually cause the interpolation matrix to be ill-conditioned. To circumvent this problem, the CS-RBF is employed to interpolate the level-set function due to its strictly positive-definite property and matrix sparseness. The Wendland's CS-RBF with C^2-smoothness is defined as

$$\varphi_i(r) = (\max(0, 1-r))^4 (4r+1) \tag{2.48}$$

Moreover, the partial derivatives of the Wendland's CS-RBF can be calculated as

$$\frac{\partial \varphi_i}{\partial x} = -20(\max(0, 1-r))^3 \frac{x-x_i}{d_p^2} \tag{2.49}$$

$$\frac{\partial \varphi_i}{\partial y} = -20(\max(0, 1-r))^3 \frac{y-y_i}{d_p^2} \tag{2.50}$$

The Wendland's CS-RBF ($x_i = y_i = 0$, $d_p = 1$) and its partial derivatives in the X and Y directions are shown in Fig. 2.28.

The level-set function $\Phi(x)$ can be defined by means of the interpolation of CS-RBF centered at n knots that are distributed in the embedding domain D.

$$\Phi(x, a) = \varphi^T(x)a = \sum_{i=1}^{n} \varphi_i(x)a_i \tag{2.51}$$

where $\varphi(x) = \{\varphi_1(x), \varphi_2(x), ..., \varphi_n(x)\}^T$ and $a = \{a_1, a_2, ..., a_n\}^T$, in which a_i is the ith unknown expansion coefficient.

2.3.3 Implicit B-spline

As in Example 2.3, the parametric function of a circle centered at $O'(x_0, y_0)$ with radius r is expressed as

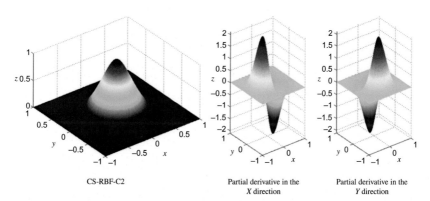

CS-RBF-C2   Partial derivative in the   Partial derivative in the
                          X direction   Y direction

FIG. 2.28 CS-RBF ($x_i = y_i = 0$, $d_p = 1$) and its partial derivatives.

$$\begin{cases} x = r\cos\theta + x_0 \\ y = r\sin\theta + y_0 \end{cases} \tag{2.52}$$

If the radius is changeable in terms of θ, the corresponding shape then becomes deformable

$$\begin{cases} x = r(\theta)\cos\theta + x_0 \\ y = r(\theta)\sin\theta + y_0 \end{cases} \tag{2.53}$$

A literature study shows that the construction of deformable models with great design flexibility (Montagnat et al., 2001) is a motivating research topic in the field of computer graphics and structural optimization. To this end, super-ellipse (Gardiner, 1965; Gridgeman, 1970) and superquadric (Barr, 1981) have been developed. The former can be seen as a generalized representation of circles, ellipses, squares, and rectangles. Zhou and Kambhamettu (2001) developed an extending superquadrics that can model complex shapes and revealed that the exponent parameter can also be changeable. Gielis (2003) and Gielis et al. (2003) introduced the superformula for the definition of generalized circles that can represent regular polygons and natural shapes with various symmetries. Notice that all the above deformable shapes can be written in parametric and implicit forms via simple transformation. Typical formulae of 2D deformable shapes are summarized in Table 2.3.

In this book, we introduce the implicit B-splines whose radius function is defined in the parametric form of B-spline curves. B-spline curves are popular in the CAD community and have been extensively used in engineering design, owing to their flexibility and local modifiability (De Boor, 1978; Piegl and Tiller, 2012). B-spline curves are parameterized in terms of intrinsic variables and expressed as the linear combination of the product of B-spline basis functions and coordinates of control points P_i ($i = 1, 2, \ldots, n$)

$$P(\xi) = \sum_{i=1}^{n} B_{i,p}(\xi) P_i \tag{2.54}$$

where ξ is the intrinsic variable with $\xi \in [0, 1]$. n denotes the number of control points. $B_{i,p}$ is the p-order B-spline basis function that holds the following recurrence relation

$$\begin{cases} B_{i,0}(\xi) = \begin{cases} 1 & \text{if } \xi_i \le \xi \le \xi_{i+1} \\ 0 & \text{otherwise} \end{cases} \\ B_{i,p}(\xi) = \dfrac{\xi - \xi_i}{\xi_{i+p} - \xi_i} B_{i,p-1}(\xi) + \dfrac{\xi_{i+p+1} - \xi}{\xi_{i+p+1} - \xi_{i+1}} B_{i+1,p-1}(\xi) \\ \text{provided} : \dfrac{0}{0} = 0 \end{cases} \tag{2.55}$$

In this book, B-spline basis functions are defined on the uniform and open knot vector $\Xi = \{\xi_1, \xi_2, \ldots, \xi_{n+p+1}\}$, which is obtained by knot refinement and degree elevation from the initial knot vector $\Xi_0 = \{0,0,1,1\}$. Fig. 2.29 illustrates

TABLE 2.3 Typical two-dimensional deformable shapes.

Shapes	Parametric expressions	Parameterization	Examples
Superquadrics (superellipse) (Gardiner, 1965; Gridgeman, 1970; Barr, 1981)	$\begin{cases} x(\theta) = a_1 \, \text{sign}(\cos\theta)\,\lvert\cos\theta\rvert^{\gamma} \\ y(\theta) = a_2 \, \text{sign}(\sin\theta)\,\lvert\sin\theta\rvert^{\gamma} \end{cases}$	a_1, a_2, and γ: designable parameters	
Extending superquadrics (Zhou and Kambhamettu, 2001)		$\gamma = \sum_{i=0}^{p} \gamma_i \overline{B}_{i,p}\left(\dfrac{\theta+\pi}{2\pi}\right), \quad \theta \in [-\pi, \pi]$ $\overline{B}_{i,p}$: Bernstein polynomials of degree p	
Gielis curves (Gielis, 2003; Gielis et al., 2003)	$\begin{cases} x(\theta) = r(\theta)\cos\theta \\ y(\theta) = r(\theta)\sin\theta \end{cases}$	$r(\theta) = \dfrac{1}{\sqrt[n_1]{\left\lvert \dfrac{1}{a}\cos\left(\dfrac{m\theta}{4}\right)\right\rvert^{n_2} + \left\lvert \dfrac{1}{b}\sin\left(\dfrac{m\theta}{4}\right)\right\rvert^{n_3}}}$	
Closed B-splines		$r(\theta) = \sum_{i=1}^{n} R_i B_{i,p}\left(\dfrac{\theta+\pi/2}{2\pi}\right), \quad \theta \in [-\pi/2,\, 3\pi/2]$ $B_{i,p}$: B-spline basis functions of degree p	

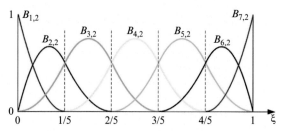

FIG. 2.29 Quadratic univariate B-spline basis functions ($p=2$, $n=7$) with the knot vector $\Xi=\{0,0,0,1/5,2/5,3/5,4/5,1,1,1\}$.

quadratic univariate B-spline basis functions defined by the knot vector $\Xi=\{0,0,0,1/5,2/5,3/5,4/5,1,1,1\}$. A detailed presentation of B-spline can be found in standard textbooks such as those of Piegl and Tiller (2012) and De Boor (1978).

The radius function of implicit B-splines is parameterized as

$$r(\theta)=\sum_{i=1}^{n}R_iB_{i,p}\left(\frac{\theta-\theta_L}{\theta_U-\theta_L}\right) \quad (2.56)$$

in which the control radius R_i corresponds to the distance between the ith control point and the prescribed center $O'(x_0, y_0)$. For any point $P=(x, y)$ in a prescribed rectangular domain $D\subset R^2$, θ corresponds to the inclined angle between vector $O'x$ and $O'P$. It is calculated as

$$\theta(x,y)=\arctan\left(\frac{y-y_0}{x-x_0}\right)+\frac{\pi}{2}\text{sign}(x-x_0)[\text{sign}(x-x_0)-1] \quad (2.57)$$

θ_L and θ_U are the minimum and maximum of θ of all points in D, satisfying that $-\pi/2\leq\theta_L\leq\theta\leq\theta_U\leq3\pi/2$.

The implicit B-spline curve in the form of the level-set function is expressed as

$$\Phi(x,y)=r(\theta)-\sqrt{(x-x_0)^2+(y-y_0)^2} \quad (2.58)$$

Note that Eq. (2.58) can describe both closed and open boundaries. A closed B-spline (CBS) is defined when $O'(x_0, y_0)$ lies inside domain D. In order to ensure the closure of $r(\theta)$ in the polar coordinate system, the same value is set to R_i related to the starting and end points. Let us now consider a set of control parameters $R=\{R_i\}=[1,3,1.5,2.5,0.5,2.5,1]^T$, with the knot vector and the B-spline basis functions shown in Fig. 2.29. In this way, a closed curve related to Eqs. (2.56)–(2.58) can be constructed in the Cartesian coordinate system. Fig. 2.30 shows a mapping relation from the Cartesian coordinate $O'\theta r$ to $O'xy$. In the coordinate system $O'\theta r$, the B-spline curve $r(\theta)$ holds the convex hull property, that is, $r(\theta)$ is enveloped within the maximum convex polygon defined

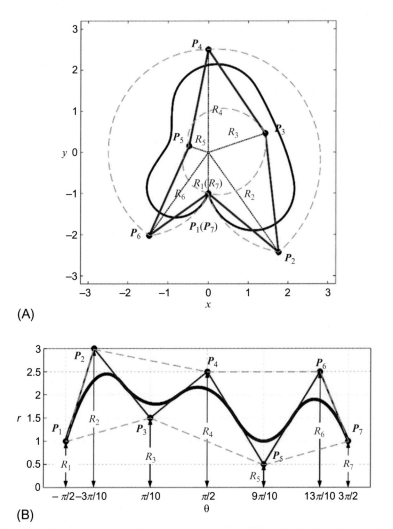

(A)

(B)

FIG. 2.30 Control points P_i and parameters R_i of a quadratic CBS: (A) in the Cartesian coordinate system $O'xy$; and (B) in the Cartesian coordinate system $O'\theta r$.

by control points $\{P_i\}$, as shown in Fig. 2.30B. Similarly, in the case of the coordinate $O'xy$ of Fig. 2.30A, the curve $r(\theta)$ is entirely inside the curve generated by the maximum convex polygon.

To avoid the poor continuity of the CBS at the end point $\theta = -\pi/2$ in Fig. 2.30, the artificial condition $P_1 = P_7 = (P_2 + P_6)/2$ or $R_1 = R_7 = (R_2 + R_6)/2$ can simply be imposed, as shown in Fig. 2.31. In this case, the CBS can be controlled by five independent design variables.

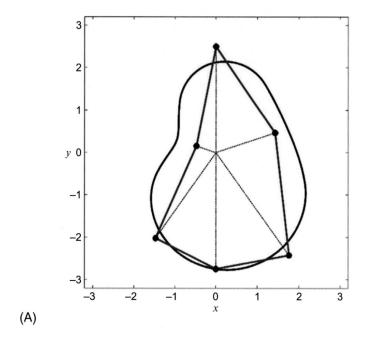

(A)

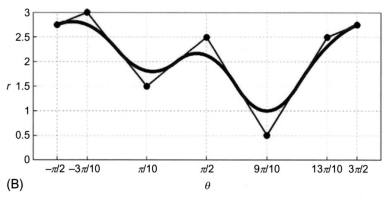

(B)

FIG. 2.31 Control points of a quadratic CBS after modifying starting point P_1 and end point P_7: (A) in the Cartesian coordinate system $O'xy$; and (B) in the Cartesian coordinate system $O'\theta r$.

An open B-spline (OBS) is defined when $O'(x_0, y_0)$ is located outside domain D. To have a clear idea, a typical example of a two-order OBS with six control points is illustrated in Fig. 2.32.

In the three-dimensional (3D) case, take the CBS as an example. Its level-set function is written as

$$\Phi(x, y, z) = r(\theta, \beta) - \sqrt{(x - x_0)^2 + (y - y_0)^2 + (z - z_0)^2} \qquad (2.59)$$

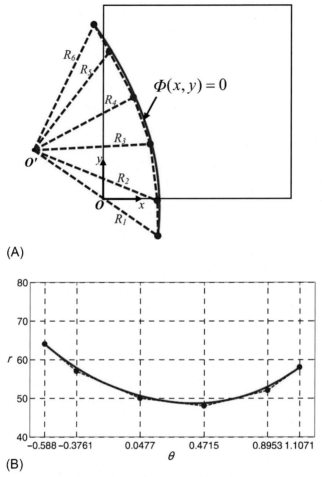

FIG. 2.32 Control points of a quadratic OBS centered at $(-30, 20)$ with six control radii $\{R_i\} = \{64, 57, 50, 48, 52, 58\}$ in domain $D = [0, 80] \times [0, 80]$: (A) in the Cartesian coordinate system $O'xy$; and (B) in the Cartesian coordinate system $O'\theta r$.

with

$$\begin{cases} r(\theta, \beta) = \sum_{i=1}^{n_1} \sum_{j=1}^{n_2} R_{i,j} B_{i,p} \left(\dfrac{\theta + \pi/2}{2\pi} \right) B_{j,q} \left(\dfrac{\beta + \pi/2}{\pi} \right) \\ \beta = \arctan \dfrac{z - z_0}{\sqrt{(x - x_0)^2 + (y - y_0)^2}}, \quad \beta \in \left[-\dfrac{\pi}{2}, \dfrac{\pi}{2} \right] \end{cases} \tag{2.60}$$

where $B_{i,p}$ and $B_{j,q}$ are two independent B-spline basis functions. As shown in Fig. 2.33, θ and β represent the inclined angles of radius r. The value of θ ranges

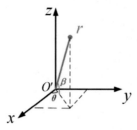

FIG. 2.33 Definition of θ and β for three-dimensional CBS.

within the interval $[-\pi/2, 3\pi/2]$ and the range of β is $[-\pi/2, \pi/2]$. In addition, the closure of the CBS requires that

$$\begin{cases} R_{1,j} = R_{n_1,j} & j = 1,2,\ldots,n_2 \\ R_{1,1} = R_{2,1} = \cdots = R_{n_1,1} \\ R_{1,n_2} = R_{2,n_2} = \cdots = R_{n_1,n_2} \end{cases} \qquad (2.61)$$

Consider a set of control parameters $R = \{R_{i,j}\} = \{4, 4, \ldots, 4\}$. Suppose two knot vectors $\varXi_1 = \{0, 0, 0, 1/9, 2/9, 3/9, 4/9, 5/9, 6/9, 7/9, 8/9, 1, 1, 1\}$ and $\varXi_2 = \{0, 0, 0, 1/7, 2/7, 3/7, 4/7, 5/7, 6/7, 1, 1, 1\}$ are used to define two two-order B-spline basis functions ($p = 2, n_1 = 11, q = 2, n_2 = 9$). Fig. 2.34A represents the distribution of control points in the Cartesian coordinate system $\theta\beta R$ and the resulting B-spline surface is shown in Fig. 2.34B. Clearly, a plane is obtained due to the same length of all control radii. Its conversion into the Cartesian coordinate xyz produces a polyhedron of control points, as illustrated in Fig. 2.34C. Especially, n_1 control points overlap at two poles. Fig. 2.34D shows the constructed CBS surface in the form of a sphere with radius $r = 4$.

2.4 Relationship between implicit and parametric functions

2.4.1 Transformation between implicit and parametric functions

An underlying relationship exists between implicit and parametric boundary representations. The conversion between implicit and parametric boundary representations theoretically constitutes a challenging topic in the community of computational geometry and computer graphics. The notion "parameterization" refers to the conversion from the implicit to parametric form, whereas "implicitization" refers to the opposite conversion from parametric to implicit form.

- **Implicitization**

In the work of Sederberg et al. (1984), it was shown that a 2D parametric curve in the form of rational polynomials of degree n can be expressed as an implicit function in the form of a polynomial of degree n. One such implicitization can be done using the elimination theory to find the existence condition for the

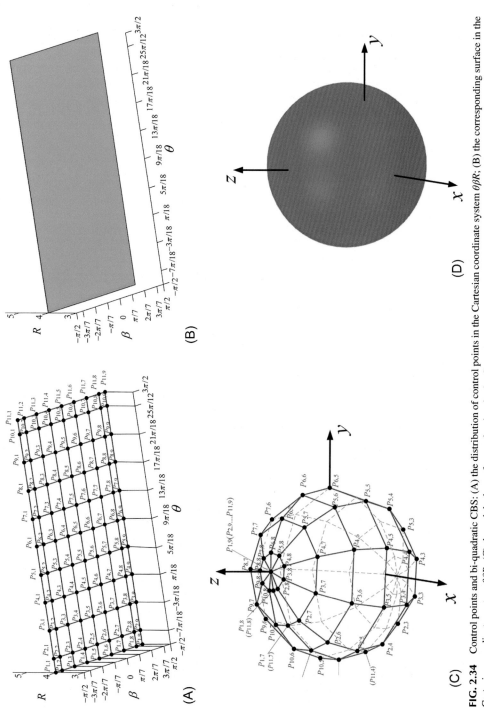

FIG. 2.34 Control points and bi-quadratic CBS: (A) the distribution of control points in the Cartesian coordinate system $\theta\beta R$; (B) the corresponding surface in the Cartesian coordinate system $\theta\beta R$; (C) the polyhedron of control points in the Cartesian coordinate system $\theta\beta R$; and (D) the corresponding surface in the Cartesian coordinate system xyz.

common roots of sets of polynomials with the determinant of Sylvester's matrix. Thus, popular CAD parametric curves such as B-splines and NURBS can be implicitized because they are either polynomials or rational polynomials.

Suppose a parametric curve is expressed as

$$
\begin{cases}
x(t) = p_m t^m + p_{m-1} t^{m-1} + \cdots + p_1 t + p_0 \\
y(t) = q_n t^n + q_{n-1} t^{n-1} + \cdots + q_1 t + q_0
\end{cases}
\tag{2.62}
$$

Eq. (2.62) can equivalently be written as

$$
\begin{cases}
0 = p_m t^m + p_{m-1} t^{m-1} + \cdots + p_1 t + (p_0 - x) \\
0 = q_n t^n + q_{n-1} t^{n-1} + \cdots + q_1 t + (q_0 - y)
\end{cases}
\tag{2.63}
$$

Besides, the following relation holds

$$
\begin{cases}
0 = p_m t^{m+n} + p_{m-1} t^{m+n-1} + \cdots + p_1 t^{n+1} + (p_0 - x) t^n \\
0 = q_n t^{m+n} + q_{n-1} t^{m+n-1} + \cdots + q_1 t^{m+1} + (q_0 - y) t^m
\end{cases}
\tag{2.64}
$$

Combing Eqs. (2.63) with (2.64), we can obtain the group of equations in the form of the matrix

$$
At = 0
\tag{2.65}
$$

where

$$
A =
\begin{bmatrix}
p_m & \cdots & p_1 & p_0 - x & 0 & 0 & \cdots & 0 \\
0 & p_m & \cdots & p_1 & p_0 - x & 0 & \cdots & 0 \\
\vdots & \vdots & \vdots & \vdots & \vdots & \vdots & \vdots & \vdots \\
0 & 0 & \cdots & 0 & p_m & \cdots & p_1 & p_0 - x \\
q_n & \cdots & q_1 & q_0 - y & 0 & 0 & \cdots & 0 \\
0 & q_n & \cdots & q_1 & q_0 - y & 0 & \cdots & 0 \\
\vdots & \vdots & \vdots & \vdots & \vdots & \vdots & \vdots & \vdots \\
0 & 0 & \cdots & 0 & q_n & \cdots & q_1 & q_0 - y
\end{bmatrix}
\tag{2.66}
$$

$$
t = \begin{bmatrix} t^{m+n} & t^{m+n-1} & \cdots & t^2 & t^1 & 1 \end{bmatrix}^T
\tag{2.67}
$$

Matrix A is Sylvester's matrix. The necessary and sufficient condition of having the nontrivial root for Eq. (2.65) is $\det A = 0$. This is the Sylvester's matrix elimination method for the implicitization of polynomial parametric curves.

Example 2.12 Implicitization of a Bezier curve

In general, a cubic Bezier curve is parametrically defined as

$$P(t) = \begin{bmatrix} 1 & t & t^2 & t^3 \end{bmatrix} \begin{bmatrix} 1 & 0 & 0 & 0 \\ -3 & 3 & 0 & 0 \\ 3 & -6 & 3 & 0 \\ -1 & 3 & -3 & 1 \end{bmatrix} \begin{bmatrix} V_0 \\ V_1 \\ V_2 \\ V_3 \end{bmatrix}, \quad 0 \le t \le 1 \qquad (2.68)$$

where V_0, V_1, V_2, V_3 are the control points of the Bezier curve.

Consider that the control points are (1, 1), (2, 2), (3, 2), and (4, 1). The corresponding parametric function of the Bezier curve is

$$\begin{cases} x = 3t + 1 \\ y = -3t^2 + 3t + 1 \end{cases}, \quad 0 \le t \le 1 \qquad (2.69)$$

The Sylvester's matrix of the above parametric function is

$$A = \begin{bmatrix} 3 & 1-x & 0 \\ 0 & 3 & 1-x \\ -3 & 3 & 1-y \end{bmatrix} \qquad (2.70)$$

Its determinant corresponds to $\det A = -3x^2 + 15x - 9y - 3$.

Thus, the implicitization of the above Bezier curve is computed as

$$\Phi(x, y) = -3x^2 + 15x - 9y - 3 = 0 \qquad (2.71)$$

Both the parametric curve and the zero-contour of the corresponding implicit function coincide completely and are illustrated in Fig. 2.35.

- **Parameterization**

In contrast, it is generally impossible to parameterize an implicit function of the polynomial of degree n. An algebraic curve has a parametric representation in

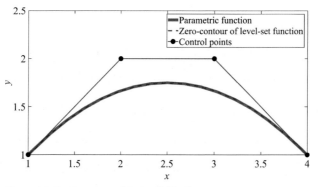

FIG. 2.35 Parametric Bezier curve and its implicitization.

the form of a rational polynomial if and only if the curve has genus zero (Walker, 1950).

2.4.2 Parametric function for the point-in-polygon test

In Section 2.1.1, it is shown that the level-set function is straightforward to identify the position of a point according to the sign of the level-set function. However, when the parametric function is used, the identification of whether a point is inside a region concerns the Jordan curve theorem. Practically, this can be carried out by the so-called point-in-polygon test (Manber, 1989; Taylor, 1994; Feito et al., 1995; Etzion and Rappoport, 1997; Huang and Shih, 1997; Teillaud, 2000; Hormann and Agathos, 2001; Rank et al., 2011; Praparata and Shamos, 1985). The polygon can be obtained by discretizing the parametric boundary related to Eqs. (2.5), (2.6).

Actually, a variety of methods are available to realize the point-in-polygon test; these can be classified into two categories. The first one, including the ray-crossing method (Manber, 1989), the triangle-based method (Feito et al., 1995), the sum-of-angles method (Hormann and Agathos, 2001), and the sign of the offset method (Taylor, 1994), directly addresses the locations of the points by computing relevant parameters. The second one is to decompose in advance the polygon into simple components such as grids (Huang and Shih, 1997), trapezoids (Teillaud, 2000), triangles (Praparata and Shamos, 1985), and star-shape polygons (Etzion and Rappoport, 1997).

Here, the ray-crossing method is introduced because of its simplicity, stability, and suitability to both convex and concave polygons. Fig. 2.36 presents a polygon defined by an array of points $P_0, P_1, \ldots, P_{n-1}$ and P_n. The basic scheme of the ray-crossing method is to draw a straight line passing point N to be identified. The line is usually parallel to one of the coordinate axes and points to the positive direction of the axis. If the number of intersections is odd, point N is inside the polygon. Otherwise, it is outside. However, it should be noted that special cases related to the intersection between the ray and the edge illustrated in Fig. 2.37 may occur and should be taken into account.

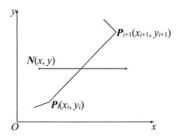

FIG. 2.36 Illustration of the ray-crossing method; $P_i P_{i+1}$ is an edge of the polygon.

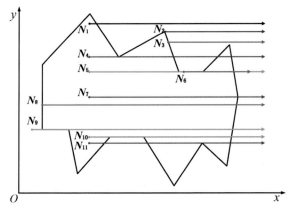

FIG. 2.37 Special cases in the ray-crossing method.

References

Allaire, G., Jouve, F., Michailidis, G., 2016. Thickness control in structural optimization via a level set method. Struct. Multidiscip. Optim. 53 (6), 1349–1382.

Andrews, L.C., 1998. Special functions of mathematics for engineers. vol. 49 SPIE Press.

Barr, A.H., 1981. Superquadrics and angle-preserving transformations. IEEE Comput. Graph. Appl. 1 (1), 11–23.

Buhmann, M.D., 2000. Radial basis functions. Acta Numer. 9, 1–38.

Carr, J.C., Beatson, R.K., Cherrie, J.B., Mitchell, T.J., Fright, W.R., McCallum, B.C., Evans, T.R., 2001. Reconstruction and representation of 3D objects with radial basis functions. In: Proceedings of the 28th Annual Conference on Computer Graphics and Interactive Techniques, pp. 67–76.

De Boor, C., 1978. A Practical Guide to Splines. Springer-Verlag, New York.

Etzion, M., Rappoport, A., 1997. On compatible star decompositions of simple polygons. IEEE Trans. Vis. Comput. Graph. 3 (1), 87–95.

Feito, F., Torres, J.C., Urena, A., 1995. Orientation, simplicity, and inclusion test for planar polygons. Comput. Graph. 19 (4), 595–600.

Fougerolle, Y.D., Gribok, A., Foufou, S., Truchetet, F., Abidi, M.A., 2005. Boolean operations with implicit and parametric representation of primitives using R-functions. IEEE Trans. Vis. Comput. Graph. 11 (5), 529–539.

Fougerolle, Y.D., Gielis, J., Truchetet, F., 2013. A robust evolutionary algorithm for the recovery of rational Gielis curves. Pattern Recogn. 46 (8), 2078–2091.

Gardiner, M., 1965. The superellipse: a curve that lies between the ellipse and the rectangle. Sci. Am. 213 (3), 222–232.

Gielis, J., 2003. A generic geometric transformation that unifies a wide range of natural and abstract shapes. Am. J. Bot. 90 (3), 333–338.

Gielis, J., Beirinckx, B., Bastiaens, E., 2003. Superquadrics with rational and irrational symmetry. In: Proceedings of the Eighth ACM Symposium on Solid Modeling and Applications, pp. 262–265.

Gridgeman, N.T., 1970. Lamé ovals. Math. Gaz. 54 (387), 31–37.

Hormann, K., Agathos, A., 2001. The point in polygon problem for arbitrary polygons. Comput. Geom. 20 (3), 131–144.

Huang, C.W., Shih, T.Y., 1997. On the complexity of point-in-polygon algorithms. Comput. Geosci. 23 (1), 109–118.

Kawamoto, A., Matsumori, T., Yamasaki, S., Nomura, T., Kondoh, T., Nishiwaki, S., 2011. Heaviside projection based topology optimization by a PDE-filtered scalar function. Struct. Multidiscip. Optim. 44 (1), 19–24.

Kumar, A.V., Padmanabhan, S., Burla, R., 2008. Implicit boundary method for finite element analysis using non-conforming mesh or grid. Int. J. Numer. Methods Eng. 74 (9), 1421–1447.

Manber, U., 1989. Introduction to Algorithms: A Creative Approach. Addison-Wesley Longman Publishing Co., Inc.

Montagnat, J., Delingette, H., Ayache, N., 2001. A review of deformable surfaces: topology, geometry and deformation. Image Vis. Comput. 19 (14), 1023–1040.

Ohtake, Y., Belyaev, A., Seidel, H.P., 2005. 3D scattered data interpolation and approximation with multilevel compactly supported RBFs. Graph. Model. 67 (3), 150–165.

Piegl, L., Tiller, W., 2012. The NURBS Book. Springer Science & Business Media.

Preparata, F.P., Shamos, M.I., 1985. Computational Geometry: An Introduction. Springer, New York.

Rank, E., Kollmannsberger, S., Sorger, C., Düster, A., 2011. Shell finite cell method: a high order fictitious domain approach for thin-walled structures. Comput. Methods Appl. Mech. Eng. 200 (45–46), 3200–3209.

Sederberg, T.W., Anderson, D.C., Goldman, R.N., 1984. Implicit representation of parametric curves and surfaces. Comput. Vis. Graph. Image Process. 28 (1), 72–84.

Skala, V., 2012. Radial basis functions for high dimensional visualization. In: VisGra-ICONS 2012, pp. 218–222.

Taubin, G., 1994. Distance approximations for rasterizing implicit curves. ACM Trans. Graph. 13 (1), 3–42.

Taylor, G., 1994. Point in polygon test. Surv. Rev. 32 (254), 479–484.

Teillaud, M., 2000. Union and split operations on dynamic trapezoidal maps. Comput. Geom. 17 (3–4), 153–163.

Walker, R.J., 1950. Algebraic Curves. Princeton University Press, Princeton, New Jersey.

Wang, M.Y., Wang, X., 2004. "Color" level sets: a multi-phase method for structural topology optimization with multiple materials. Comput. Methods Appl. Mech. Eng. 193 (6–8), 469–496.

Wang, M.Y., Wang, X., Guo, D., 2003. A level set method for structural topology optimization. Comput. Methods Appl. Mech. Eng. 192 (1–2), 227–246.

Xia, L., Zhu, J., Zhang, W., Breitkopf, P., 2013. An implicit model for the integrated optimization of component layout and structure topology. Comput. Methods Appl. Mech. Eng. 257, 87–102.

Zhang, J., Zhang, W.H., Zhu, J.H., Xia, L., 2012. Integrated layout design of multi-component systems using XFEM and analytical sensitivity analysis. Comput. Methods Appl. Mech. Eng. 245, 75–89.

Zhao, H.K., Chan, T., Merriman, B., Osher, S., 1996. A variational level set approach to multiphase motion. J. Comput. Phys. 127 (1), 179–195.

Zhou, L., Kambhamettu, C., 2001. Extending superquadrics with exponent functions: modeling and reconstruction. Graph. Models 63 (1), 1–20.

Chapter 3

Basic operations of level-set functions

3.1 Operations of a single level-set function

3.1.1 Translation, rotation, and scaling operations

In this chapter, the basic geometric and topological operations of shape features are introduced in terms of level-set functions. First, the translation, rotation, scaling, and a combination of them are discussed. These operations constitute the basic operations or transformations in computer-aided geometry design (CAGD).

- **Translation**

For a feature defined by the level-set function $\Phi(x) \geq 0$, the translation with quantity $\triangle x$ changes the level-set function into

$$\Phi(x - \Delta x) \geq 0 \tag{3.1}$$

- **Rotation**

As is well known, rotation can be accomplished by multiplying the coordinates by a rotation matrix R. The level-set function of the concerned feature after rotation can thus be written as

$$\Phi(Rx) \geq 0 \tag{3.2}$$

Rotation in space is classified into three cases in accordance with the rotation axis: rotations around the x-axis, y-axis, and z-axis. The rotation matrices R are expressed as

$$R(\alpha) = \begin{bmatrix} 1 & 0 & 0 \\ 0 & \cos\alpha & \sin\alpha \\ 0 & -\sin\alpha & \cos\alpha \end{bmatrix}, \text{ Rotation around } x-\text{axis} \tag{3.3}$$

The Feature-driven Method for Structural Optimization. https://doi.org/10.1016/B978-0-12-821330-8.00003-1
© 2021 Shanghai Jiao Tong University Press. Published by Elsevier Inc. All rights reserved.

$$\boldsymbol{R}(\beta) = \begin{bmatrix} \cos\beta & 0 & -\sin\beta \\ 0 & 1 & 0 \\ \sin\beta & 0 & \cos\beta \end{bmatrix}, \quad \text{Rotation around } y-\text{axis} \quad (3.4)$$

$$\boldsymbol{R}(\gamma) = \begin{bmatrix} \cos\gamma & \sin\gamma & 0 \\ -\sin\gamma & \cos\gamma & 0 \\ 0 & 0 & 1 \end{bmatrix}, \quad \text{Rotation around } z-\text{axis} \quad (3.5)$$

where α, β, and γ denote the anticlockwise rotation angles relative to the x-axis, y-axis, and z-axis, respectively.

Besides, the combination of rotation around different axes can be realized by multiplying the rotation matrices in Eqs. (3.3)–(3.5). Suppose a feature successively rotates around the x-axis, y-axis, and z-axis. The combined rotation matrix then corresponds to

$$\boldsymbol{R}_{xyz} = \boldsymbol{R}(\gamma)\boldsymbol{R}(\beta)\boldsymbol{R}(\alpha) \quad (3.6)$$

The above matrix multiplication does not satisfy the commutative law. Six different rotation matrices can thus be obtained for different rotation sequences, respectively. They are $\boldsymbol{R}_{xyz}$, $\boldsymbol{R}_{xzy}$, $\boldsymbol{R}_{yxz}$, $\boldsymbol{R}_{yzx}$, $\boldsymbol{R}_{zxy}$, and $\boldsymbol{R}_{zyx}$.

• **Scaling**

Suppose the scaling of a feature is made along a certain direction x_i. The corresponding level-set function is then expressed as

$$\Phi\left(x_1, x_2, \ldots, \frac{1}{m_i}x_i, \ldots, , x_n\right) \geq 0 \quad (3.7)$$

where m_i is the scaling factor. The feature is enlarged along the direction of x_i when $m_i > 1$ and shrunk when $m_i < 1$.

Example 3.1 Translation, rotation, and scaling of a superellipse
To have a clear idea, consider a superellipse whose level-set function is expressed as

$$\Phi(x, y) = 1 - \left|\frac{x}{l}\right|^m - \left|\frac{y}{w}\right|^m \quad (3.8)$$

where $m = 50$. Its semilength and semiwidth are $l = 4$ and $w = 1$, respectively.

Suppose the superellipse is translated with $\Delta x = 3$ along the x-axis and $\Delta y = -4$ along the y-axis. The level-set function is then changed into

$$\Phi(x, y) = 1 - \left|\frac{x - \Delta x}{l}\right|^m - \left|\frac{y - \Delta y}{w}\right|^m \quad (3.9)$$

Suppose the superellipse rotates with an angle $\theta = \pi/4$ anticlockwise. The level-set function thus becomes

$$\Phi(x, y) = 1 - \left|\frac{x\cos\theta + y\sin\theta}{l}\right|^m - \left|\frac{-x\sin\theta + y\cos\theta}{w}\right|^m \quad (3.10)$$

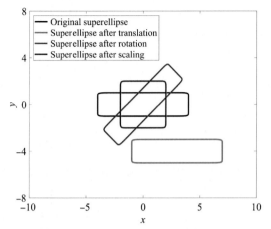

FIG. 3.1 Translation, rotation, and scaling of superellipse in terms of level-set function.

Besides, suppose it is scaled with factors $m_1 = 1/2$ along the x-axis and $m_2 = 2$ along the y-axis. The scaling operation then results in

$$\Phi(x, y) = 1 - \left|\frac{1}{m_1} \cdot \frac{x}{l}\right|^m - \left|\frac{1}{m_2} \cdot \frac{y}{w}\right|^m \quad (3.11)$$

Fig. 3.1 shows the translation, rotation, and scaling of the superellipse by the zero-contours of the above level-set functions.

Example 3.2 Translation, rotation, and scaling of a T-shape feature
Consider a T-shape feature defined with the level-set function $\Phi(x,y,z) \geq 0$. The translation, rotation, and scaling operations act on the T-shape feature successively.

First, the feature is scaled with scaling factors $m_1 = m_2 = m_3 = 2$ along the x-axis, y-axis, and z-axis, respectively. The level-set function of the feature after scaling is thus written as

$$\Phi\left(\frac{x}{m_1}, \frac{y}{m_2}, \frac{z}{m_3}\right) \geq 0 \quad (3.12)$$

Second, the feature after scaling is translated with $\Delta x = 2$ along the x-axis, $\Delta y = 1$ along the y-axis, and $\Delta z = 2$ along the z-axis, respectively. The level-set function is then changed into

$$\Phi\left(\frac{x}{m_1} - \Delta x, \frac{y}{m_2} - \Delta y, \frac{z}{m_3} - \Delta z\right) \geq 0 \quad (3.13)$$

The feature after scaling and translation is further rotated with $\gamma = 3\pi/2$ around the z-axis, $\beta = \pi/4$ around the y-axis, and $\alpha = \pi/4$ around the x-axis, successively. The following level-set function is then obtained

$$\Phi\left(\mathbf{R}_{zyx}\bar{\mathbf{x}}\right) \geq 0 \quad (3.14)$$

in which the rotation matrix $\boldsymbol{R}_{zyx}$ and transformed coordinates $\bar{\boldsymbol{x}}$ are expressed as

$$\boldsymbol{R}_{zyx} = \boldsymbol{R}(\alpha)\boldsymbol{R}(\beta)\boldsymbol{R}(\gamma)$$

$$= \begin{bmatrix} 1 & 0 & 0 \\ 0 & \cos\alpha & \sin\alpha \\ 0 & -\sin\alpha & \cos\alpha \end{bmatrix} \begin{bmatrix} \cos\beta & 0 & -\sin\beta \\ 0 & 1 & 0 \\ \sin\beta & 0 & \cos\beta \end{bmatrix} \begin{bmatrix} \cos\gamma & \sin\gamma & 0 \\ -\sin\gamma & \cos\gamma & 0 \\ 0 & 0 & 1 \end{bmatrix} \quad (3.15)$$

$$\bar{\boldsymbol{x}} = [x/m_1 - \Delta x \quad y/m_2 - \Delta y \quad z/m_3 - \Delta z]^{\mathrm{T}} \quad (3.16)$$

The original and transformed T-shape features are shown in Fig. 3.2 by the zero-isosurfaces of the above level-set functions.

3.1.2 Twisting, sweeping, and polynomial operations

More complex operations can also be applied to a feature defined with level-set function, such as twisting, sweeping, and polynomial operations.

- **Twisting**

Twisting of a feature around the x-axis, y-axis, and z-axis can be realized by applying the following transformations on coordinates (Reiner et al., 2011).

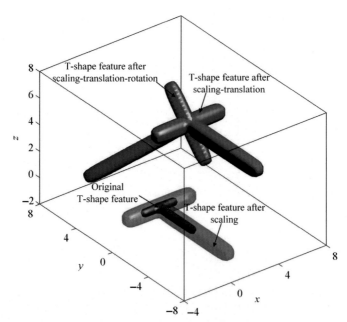

FIG. 3.2 Translation, rotation, and scaling of T-shape feature in terms of level-set function.

$$\boldsymbol{Tx} = \begin{bmatrix} x \\ y\cos(ax+b) - z\sin(ax+b) \\ y\sin(ax+b) + z\cos(ax+b) \end{bmatrix} \quad \text{Twisting aroud } x-\text{axis} \quad (3.17)$$

$$\boldsymbol{Tx} = \begin{bmatrix} x\cos(ay+b) - z\sin(ay+b) \\ y \\ x\sin(ay+b) + z\cos(ay+b) \end{bmatrix} \quad \text{Twisting aroud } y-\text{axis} \quad (3.18)$$

$$\boldsymbol{Tx} = \begin{bmatrix} x\cos(az+b) - y\sin(az+b) \\ x\sin(az+b) + y\cos(az+b) \\ z \end{bmatrix} \quad \text{Twisting aroud } z-\text{axis} \quad (3.19)$$

Example 3.3 Twisting of a cuboid
To make it clear, consider a cuboid defined with the level-set function

$$\Phi(x,y,z) = \min(\Phi_1, \Phi_2, \Phi_3) \geq 0 \qquad (3.20)$$

where

$$\begin{cases} \Phi_1(x,y,z) = 36 - x^2 \\ \Phi_2(x,y,z) = 1 - y^2 \\ \Phi_3(x,y,z) = 1 - z^2 \end{cases} \qquad (3.21)$$

By twisting the cuboid around the x-axis, the level-set function is changed into

$$\Phi(x,y,z) = \min(\Phi_1, \Phi_2, \Phi_3) \geq 0 \qquad (3.22)$$

where

$$\begin{cases} \Phi_1(x,y,z) = 36 - x^2 \\ \Phi_2(x,y,z) = 1 - [y\cos(ax+b) - z\sin(ax+b)]^2 \\ \Phi_3(x,y,z) = 1 - [y\sin(ax+b) + z\cos(ax+b)]^2 \end{cases} \qquad (3.23)$$

Fig. 3.3 shows the twisting deformation of the cuboid around the x-axis with $a=0.8$ and $b=-0.5$.

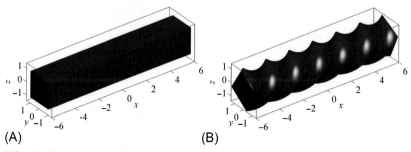

(A)   (B)

FIG. 3.3 Twisting of cuboid around x-axis: (A) original cuboid; and (B) cuboid after twisting.

Example 3.4 Twisting of a tri-cylinder

Consider a tri-cylinder as shown in Fig. 3.4A. Its level-set function is described as

$$\Phi(x, y, z) = \max\left(\Phi_1, \Phi_2, \Phi_3, \Phi_4\right) \geq 0 \tag{3.24}$$

where

$$\begin{cases} \Phi_1(x, y, z) = 36 - z^2 \\ \Phi_2(x, y, z) = 1 - \sqrt{x^2 + (y+1)^2} \\ \Phi_3(x, y, z) = 1 - \sqrt{\left(x - \sqrt{3}/2\right)^2 + (y - 1/2)^2} \\ \Phi_4(x, y, z) = 1 - \sqrt{\left(x + \sqrt{3}/2\right)^2 + (y - 1/2)^2} \end{cases} \tag{3.25}$$

Twisting the cuboid around the z-axis, the level-set function is changed into

$$\Phi(x, y, z) = \max\left(\Phi_1, \Phi_2, \Phi_3, \Phi_4\right) \geq 0 \tag{3.26}$$

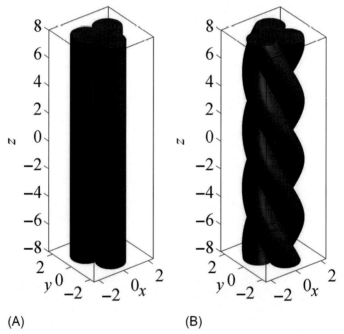

(A) (B)

FIG. 3.4 Twisting of tri-cylinder around z-axis: (A) original tri-cylinder; and (B) tri-cylinder after twisting.

where

$$
\begin{cases}
\Phi_1(x,y,z) = 36 - z^2 \\
\Phi_2(x,y,z) = 1 - \sqrt{[x\cos(az+b) - y\sin(az+b)]^2 + [x\cos(az+b) - y\sin(az+b) + 1]^2} \\
\Phi_3(x,y,z) = 1 - \sqrt{\left(x\cos(az+b) - y\sin(az+b) - \sqrt{3}/2\right)^2 + (x\cos(az+b) - y\sin(az+b) - 1/2)^2} \\
\Phi_4(x,y,z) = 1 - \sqrt{\left(x\cos(az+b) - y\sin(az+b) + \sqrt{3}/2\right)^2 + (x\cos(az+b) - y\sin(az+b) - 1/2)^2}
\end{cases}
$$

$$(3.27)$$

Fig. 3.4B shows the tri-cylinder after twisting deformation around the z-axis with $a=0.5$ and $b=-0.4$.

- **Sweeping**

Sweeping is one such operation that can be used to generate a feature from its initial cross-section along a specific trajectory. Suppose the cross-section is defined by the level-set function $\Phi(x) \geq 0$. The sweeping trajectory is defined by

$$
Sx = \begin{bmatrix} x \\ f(x) \\ g(x) \end{bmatrix}, \quad \text{Sweeping along } x - \text{axis} \tag{3.28}
$$

$$
Sx = \begin{bmatrix} f(y) \\ y \\ g(y) \end{bmatrix}, \quad \text{Sweeping along } y - \text{axis} \tag{3.29}
$$

$$
Sx = \begin{bmatrix} f(z) \\ g(z) \\ z \end{bmatrix}, \quad \text{Sweeping along } z - \text{axis} \tag{3.30}
$$

The sweeping feature can be described with the following level-set function

$$\Phi(x - Sx) \geq 0 \tag{3.31}$$

Example 3.5 Sweeping from cross-sections of a circle and a rectangle
Consider a feature obtained by sweeping from a circular cross-section along the z-axis. The level-set function related to the circular cross-section centered in plane Oxy with radius r is expressed as

$$\Phi(x,y) = r^2 - x^2 - y^2 \tag{3.32}$$

The sweeping trajectory is a helix curve that is stated as

$$
Sx = \begin{bmatrix} \cos(z)/2 \\ \sin(z)/2 \\ z \end{bmatrix} \tag{3.33}
$$

Then, the level-set function of this sweeping feature can be written as

$$\Phi(x, y, z) = r^2 - [x - \cos(z)/2]^2 - [y - \sin(z)/2]^2 \qquad (3.34)$$

Another feature is obtained by sweeping from a rectangular cross-section along the sweeping trajectory that is defined as

$$Sx = \begin{bmatrix} x \\ \sin(x)/2 \\ \sin(x)/2 \end{bmatrix} \qquad (3.35)$$

The level-set function of this feature is expressed as

$$\Phi(x, y, z) = \min(\Phi_1, \Phi_2) \qquad (3.36)$$

where

$$\begin{cases} \Phi_1(x, y, z) = l^2 - [y - \sin(x)/2]^2 \\ \Phi_2(x, y, z) = w^2 - [z - \sin(x)/2]^2 \end{cases} \qquad (3.37)$$

Fig. 3.5 shows the results of sweeping operations from the circular and rectangular cross-sections, respectively. The *black lines* illustrate the sweeping trajectories.

- **Polynomial operations**

Polynomial transformation can be regarded as a nonlinear scaling operation. For an arbitrary coordinate x_i, one such transformation corresponds to

$$x_i = a_{i,0} + a_{i,1}x_i + a_{i,2}x_i^2 + a_{i,3}x_i^3 + \cdots + a_{i,m}x_i^m \qquad (3.38)$$

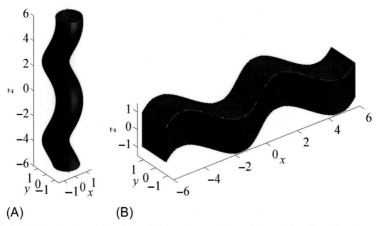

(A) (B)

FIG. 3.5 Sweeping operations from different cross-sections: (A) sweeping from circular cross-section; and (B) sweeping from rectangular cross-section.

where the polynomial order m and the involved coefficients $a_{i,0}$, $a_{i,1}$, ..., $a_{i,m}$ determine the scaling effects.

Example 3.6 Quadratic transformation of a solid sphere and a torus
To have an idea, consider the quadratic transformation $x \rightarrow x^2$ of the two features in Fig. 3.6. The level-set function of the solid sphere with radius R is stated as

$$\Phi(x, y, z) = R^2 - x^2 - y^2 - z^2 \tag{3.39}$$

After the quadratic transformation, the level-set function is changed into

$$\Phi(x, y, z) = R^2 - x^4 - y^2 - z^2 \tag{3.40}$$

The level-set function of the torus is expressed as

$$\Phi(x, y, z) = \left(x^2 + y^2 + z^2 + R^2 - r^2\right)^2 - 4R^2\left(x^2 + y^2\right) \tag{3.41}$$

Similarly, the quadratic transformation changes the level-set function into

$$\Phi(x, y, z) = \left(x^4 + y^2 + z^2 + R^2 - r^2\right)^2 - 4R^2\left(x^4 + y^2\right) \tag{3.42}$$

Fig. 3.7 shows the transformed features. Clearly, the quadratic transformation $x \rightarrow x^2$ has the effect of compressing the object along the x-axis.

3.2   Operations of multiple level-set functions

The above presentations indicate that the operations of the single level-set function provide great flexibility in changing the geometric shape and position of a feature. Especially, proper combinations of these basic operations have miraculous functionalities for the simplicity and efficiency of geometric modeling.

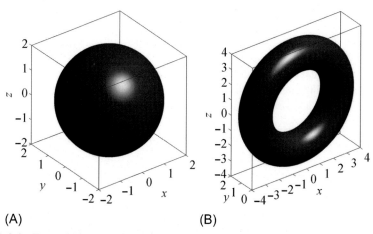

(A)   (B)

FIG. 3.6   Two solid features: (A) a sphere; and (B) a torus.

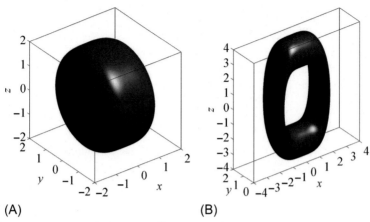

FIG. 3.7 Quadratic transformation $x \rightarrow x^2$ of two solid features: (A) sphere after quadratic transformation; and (B) torus after quadratic transformation.

Here, operations of multiple level-set functions are presented to highlight the importance in feature driven engineering and artist designs.

3.2.1 Blending operation

A blending operation is a weighting sum of multiple features into a complex object (Pasko et al., 2005). The blending of two features defined with level-set functions $\Phi_1(x)$ and $\Phi_2(x)$ corresponds to the following expression

$$\Phi(x) = \alpha\Phi_1(x) + (1-\alpha)\Phi_2(x), \quad 0 \leq \alpha \leq 1 \tag{3.43}$$

where α is the blending parameter. The blending operation can also be interpreted as the interpolation of two zero-value surfaces.

Example 3.7 Blending operation between a circle and a square, and a sphere and a cylinder

Two-dimensional (2D) and three-dimensional (3D) examples are given to illustrate the blending operation. In the 2D case, the blending of a circle and a square is studied. The level-set functions of two features are expressed as

$$\begin{cases} \Phi_1(x, y) = R^2 - x^2 - y^2 \\ \Phi_2(x, y) = (R^2 - x^2) + (R^2 - y^2) - \sqrt{(R^2 - x^2)^2 + (R^2 - y^2)^2} \end{cases} \tag{3.44}$$

As shown in Fig. 3.8A, the blending result falls between the circle and the square. α controls the blending effect. The obtained feature approaches the square when $\alpha \rightarrow 0$ and the circle when $\alpha \rightarrow 1$.

Besides, the blending between a solid sphere and a cylinder is considered. The level-set functions of the sphere and the cylinder are expressed as

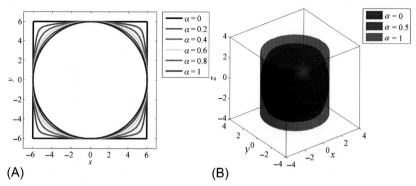

FIG. 3.8 Blending operations of two features: (A) blending of circle and square; and (B) blending of sphere and cylinder.

$$\begin{cases} \Phi_1(x, y, z) = R^2 - x^2 - y^2 - z^2 \\ \Phi_2(x, y, z) = (R^2 - x^2 - y^2) + (16 - z^2) - \sqrt{(R^2 - x^2 - y^2)^2 + (16 - z^2)^2} \end{cases}$$

(3.45)

The blending results of the sphere and the cylinder with different values of α are shown in Fig. 3.8B.

3.2.2 Boolean operations of features

In engineering design practice, constructive solid geometry (CSG) is a practical geometry modeling technique. The basic scheme of CSG is to construct an object of complex geometry with simple primitives through Boolean operations consisting of intersection, union, and difference (Requicha and Voelcker, 1977). This framework is illustrated in Fig. 3.9. The geometry information is stored in the form of tree structures. Each leaf stores a simple primitive (e.g., a circle and a rectangle in a plane or sphere and a cuboid in space) while the branch represents the corresponding Boolean operation. The root is finally the target geometry. Here, the symbols ∩ and ∪ denote the Boolean intersection and union between two primitives, respectively. Notice that the Boolean difference is also an intersection but applied between one primitive and the complement of the other primitive.

From the viewpoint of structural optimization, each simple primitive can be considered a basic design feature whose level-set function is easily formulated. The procedure of structure design such as topology optimization is then transformed into the procedure of Boolean operations involving design variables to control the position and deformation of each featured primitive.

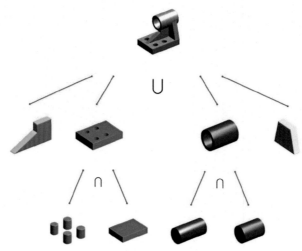

FIG. 3.9   CSG modeling of a mechanical part with Boolean operations.

Table 3.1 lists 16 different realizations in terms of two simple featured primitives, that is, a square solid and a circular solid defined by the following level-set functions.

$$\begin{cases} \Phi_A(x,y) = (10-|x|) \wedge (10-|y|) \\ \Phi_B(x,y) = 10 - \sqrt{(x-6)^2 + (y-6)^2} \end{cases} \tag{3.46}$$

Notice that the symbols $\vee$ and $\wedge$ denote the operators of Boolean union and intersection in terms of level-set functions. In Table 3.1, the complements of both featured primitives are indicated by A^C and B^C (see No. 2 and No. 10), representing the infinite solid region cut by the feature. No. 3 and No. 11 represent the infinite solid and void, respectively.

3.2.3   Boolean operations of features with max and min functions

Mathematically, it can be proved that the Boolean union and intersection operations can be realized by max and min functions.

As illustrated in Fig. 3.10A, a domain Ω is obtained as a result of the Boolean union of n subdomains $\Omega_1, \Omega_2, \ldots, \Omega_n$ represented by the level-set functions $\{\Phi_i(x) \geq 0 | i=1, 2, \ldots, n\}$. For an arbitrary point x outside domain Ω, we have $\{\Phi_i(x) \leq 0 | i=1, 2, \ldots, n\}$. Choose a contour line outside domain Ω as the zero-level contour of function Φ with $\Phi(x)=0$. Thus, Φ satisfies $\{\Phi \geq \Phi_i(x) | i=1, 2, \ldots, n\}$ on the contour line. The contour line can be described as the point set satisfying

$$A = \left\{ x \middle| \Phi(x) = 0, \Phi(x) \geq \max_{i=1,\ldots,n} \Phi_i(x) \right\} \tag{3.47}$$

TABLE 3.1 Sixteen Boolean operations between a square and a circular solid in terms of level-set functions.

No.	Geometric domain and its LSF Φ	Diagram	No.	Geometric domain and its LSF Φ	Diagram
1	A $\Phi = \Phi_A$		9	B $\Phi = \Phi_B$	
2	A^C $\Phi = -\Phi_A$		10	B^C $\Phi = -\Phi_B$	
3	$A \cup A^C$ or $B \cup B^C$ $\Phi = \lvert\Phi_A\rvert$ or $\Phi = \lvert\Phi_B\rvert$		11	$A \cap A^C$ or $B \cap B^C$ $\Phi = -\lvert\Phi_A\rvert$ or $\Phi = -\lvert\Phi_B\rvert$	
4	$A \cup B$ $\Phi = \Phi_A \vee \Phi_B$		12	$A \cap B$ $\Phi = \Phi_A \wedge \Phi_B$	
5	$A^C \cup B^C$ $\Phi = (-\Phi_A) \vee (-\Phi_B)$		13	$A^C \cap B^C$ $\Phi = (-\Phi_A) \wedge (-\Phi_B)$	
6	$A \cup B^C$ $\Phi = \Phi_A \vee (-\Phi_B)$		14	$A \cap B^C$ $\Phi = \Phi_A \wedge (-\Phi_B)$	
7	$A^C \cup B$ $\Phi = (-\Phi_A) \vee \Phi_B$		15	$A^C \cap B$ $\Phi = (-\Phi_A) \wedge \Phi_B$	
8	$(A \cap B^C) \cup (A^C \cap B)$ $\Phi = (\Phi_A \wedge (-\Phi_B)) \vee ((-\Phi_A) \wedge \Phi_B)$		16	$(A \cap B) \cup (A^C \cap B^C)$ $\Phi = (\Phi_A \wedge \Phi_B) \vee ((-\Phi_A) \wedge (-\Phi_B))$	

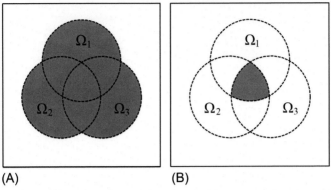

FIG. 3.10 Boolean union and intersection of n subdomains Ω_1, Ω_2, ..., Ω_n: (A) Boolean union; and (B) Boolean intersection.

The infimum of set A is the max function of $\{\Phi_i(x)|i=1, 2, ..., n\}$ and can be written as

$$\inf(A) = \left\{x\Big|\ \Phi(x) = \max_{i-1,...,n} \Phi_i(x) = 0\right\} \tag{3.48}$$

On the other hand, the infimum of set A geometrically represents the boundary of the Boolean union domain Ω. It means that the max function is the level-set function of the Boolean union operation.

A similar conclusion can be obtained for the Boolean intersection domain in Fig. 3.10B. For an arbitrary point x inside domain Ω, we have $\{\Phi_i(x) \geq 0 | i=1, 2, ..., n\}$. Choose a contour line inside domain Ω as the zero-level contour of function Φ with $\Phi(x)=0$. Thus, Φ satisfies $\{\Phi \leq \Phi_i(x) | i=1, 2, ..., n\}$ on the contour line. The contour line can be described as the point set satisfying

$$A = \left\{x\Big|\ \Phi(x) = 0, \Phi(x) \leq \min_{i=1,...,n} \Phi_i(x)\right\} \tag{3.49}$$

The supremum of set A is the min function of $\{\Phi_i(x)|i=1, 2, ..., n\}$ and can be written as

$$\sup(A) = \left\{x\Big|\ \Phi(x) = \min_{i=1,...,n} \Phi_i(x) = 0\right\} \tag{3.50}$$

Thus, it concludes that the min function is the level-set function of the Boolean intersection domain.

3.3 Typical max and min functions

Variants and approximations of max and min functions can be used to realize Boolean operations on level-set functions. Below is a detailed presentation of typical max and min functions including R-function (Rvachev, 1982; Shapiro, 1991), Ricci function (Ricci, 1973), KS function (Kreisselmeier and Steinhauser, 1978), and step function (Kumar et al., 2008). Their differentiability and regularity are studied. A summarization of these functions is listed in Table 3.2.

3.3.1 R-function

Suppose Φ_1 and Φ_2 are the level-set functions related to two featured primitives. A popular definition of the R_α function corresponds to (Shapiro and Tsukanov, 1999; Shapiro, 2007)

$$\begin{cases} \Phi_1 \vee_\alpha \Phi_2 = \dfrac{1}{1+\alpha}\left(\Phi_1 + \Phi_2 + \sqrt{\Phi_1^2 + \Phi_2^2 - 2\alpha\Phi_1\Phi_2}\right) \\ \Phi_1 \wedge_\alpha \Phi_2 = \dfrac{1}{1+\alpha}\left(\Phi_1 + \Phi_2 - \sqrt{\Phi_1^2 + \Phi_2^2 - 2\alpha\Phi_1\Phi_2}\right) \end{cases} \tag{3.51}$$

in which α is an arbitrary symmetric function that satisfies $-1 < \alpha \le 1$. When $\alpha = 0$, the most-commonly used R_0 function is obtained

$$\begin{cases} \Phi_1 \vee_0 \Phi_2 = \Phi_1 + \Phi_2 + \sqrt{\Phi_1^2 + \Phi_2^2} \\ \Phi_1 \wedge_0 \Phi_2 = \Phi_1 + \Phi_2 - \sqrt{\Phi_1^2 + \Phi_2^2} \end{cases} \tag{3.52}$$

When $\alpha = 1$, it follows that

$$\begin{cases} \Phi_1 \vee_1 \Phi_2 = \dfrac{1}{2}\left(\Phi_1 + \Phi_2 + \sqrt{\Phi_1^2 + \Phi_2^2 - 2\Phi_1\Phi_2}\right) = \max(\Phi_1, \Phi_2) \\ \Phi_1 \wedge_1 \Phi_2 = \dfrac{1}{2}\left(\Phi_1 + \Phi_2 - \sqrt{\Phi_1^2 + \Phi_2^2 - 2\Phi_1\Phi_2}\right) = \min(\Phi_1, \Phi_2) \end{cases} \tag{3.53}$$

It is noted that the max and min functions are special cases of the R_α function when $\alpha = 1$. The max and min functions are not differentiable at the points $\{x \,|\, \Phi_1(x) = \Phi_2(x)\}$ while the set of nondifferentiable points is reduced to $\{x \,|\, \Phi_1(x) = \Phi_2(x) = 0\}$ for other R_α functions. Besides, the max and min functions are fully normalized along $\Phi_1(x) = 0$ and $\Phi_2(x) = 0$ while the R_0 function is normalized into the first order along $\Phi_1(x) = 0$ or $\Phi_2(x) = 0$.

The R_0^m function is m-times differentiable. It is stated as

$$\begin{cases} \Phi_1 \vee_0^m \Phi_2 = \left(\Phi_1 + \Phi_2 + \sqrt{\Phi_1^2 + \Phi_2^2}\right)\left(\Phi_1^2 + \Phi_2^2\right)^{\frac{m}{2}} \\ \Phi_1 \wedge_0^m \Phi_2 = \left(\Phi_1 + \Phi_2 - \sqrt{\Phi_1^2 + \Phi_2^2}\right)\left(\Phi_1^2 + \Phi_2^2\right)^{\frac{m}{2}} \end{cases} \tag{3.54}$$

where m is an integer. In contrast to the R_0 function, this function is not normalized.

TABLE 3.2 Typical max and min functions corresponding to Boolean union and intersection.

Boolean operation			Union	Intersection	Differentiability	Regularity
Venn diagram						
Mathematical formulation	Level-set function	Maximum/minimum function	$\Omega_1 \cup \Omega_2 \cup \cdots \cup \Omega_n$ $\Phi = \max(\Phi_1, \Phi_2, \ldots, \Phi_n)$	$\Omega_1 \cap \Omega_2 \cap \cdots \cap \Omega_n$ $\Phi = \min(\Phi_1, \Phi_2, \ldots, \Phi_n)$	Not differentiable at $\Phi_1 = \Phi_2$	Fully normalized
		R-function: R_α function $(-1 < \alpha < 1)$	$\Phi = \Phi_1 \vee_\alpha \Phi_2 \vee_\alpha \cdots \vee_\alpha \Phi_n$ $\Phi_i \vee_\alpha \Phi_j = \frac{1}{1+\alpha}(\Phi_i + \Phi_j) + \sqrt{\Phi_i^2 + \Phi_j^2 - 2\alpha\Phi_i\Phi_j}$	$\Phi = \Phi_1 \wedge_\alpha \Phi_2 \wedge_\alpha \cdots \wedge_\alpha \Phi_n$ $\Phi_i \wedge_\alpha \Phi_j = \frac{1}{1+\alpha}(\Phi_i + \Phi_j) - \sqrt{\Phi_i^2 + \Phi_j^2 - 2\alpha\Phi_i\Phi_j}$	Not differentiable at $\Phi_1 = \Phi_2 = 0$	Normalized to order 1 when $\alpha = 0$
		R_0^m function	$\Phi = \Phi_1 \vee_0^m \Phi_2 \vee_0^m \cdots \vee_0^m \Phi_n$ $\Phi_i \vee_0^m \Phi_j = (\Phi_i + \Phi_j) + \sqrt{\Phi_i^2 + \Phi_j^2}\left(\Phi_i^2 + \Phi_j^2\right)^{\frac{m}{2}}$	$\Phi = \Phi_1 \wedge_0^m \Phi_2 \wedge_0^m \cdots \wedge_0^m \Phi_n$ $\Phi_i \wedge_0^m \Phi_j = (\Phi_i + \Phi_j) - \sqrt{\Phi_i^2 + \Phi_j^2}\left(\Phi_i^2 + \Phi_j^2\right)^{\frac{m}{2}}$	m-times differentiable	Not normalized
		R_p function	$\Phi = \Phi_1 \vee_p \Phi_2 \vee_p \cdots \vee_p \Phi_n$ $\Phi_i \vee_p \Phi_j = \Phi_i + \Phi_j + \left(\Phi_i^p + \Phi_j^p\right)^{\frac{1}{p}}$	$\Phi = \Phi_1 \wedge_p \Phi_2 \wedge_p \cdots \wedge_p \Phi_n$ $\Phi_i \wedge_p \Phi_j = \Phi_i + \Phi_j - \left(\Phi_i^p + \Phi_j^p\right)^{\frac{1}{p}}$	Not differentiable at $\Phi_1 = \Phi_2 = 0$	Normalized to order $p-1$

KS function	$\Phi = \dfrac{1}{p} \ln\left(\displaystyle\sum_{i=1}^{n} e^{p(\Phi_i - \Phi_{max})}\right) + \Phi_{max}, \ p > 0$ $\Phi_{max} = \max(\Phi_1, \Phi_2, \ldots, \Phi_n)$	$\Phi = -\dfrac{1}{p} \ln\left(\displaystyle\sum_{i=1}^{n} e^{-p(\Phi_i - \Phi_{min})}\right) + \Phi_{min}, \ p > 0$ $\Phi_{min} = \min(\Phi_1, \Phi_2, \ldots, \Phi_n)$	Differentiable everywhere	Bounded normalized
Ricci function	$\Phi = \left(\displaystyle\sum_{i=1}^{n} \Phi_i^{p}\right)^{\frac{1}{p}}, \ p > 0$	$\Phi = \left(\displaystyle\sum_{i=1}^{n} \Phi_i^{-p}\right)^{-\frac{1}{p}}, \ p > 0$	Differentiable everywhere	Bounded normalized
Step function	$\Phi = \Phi_1 \vee \Phi_2 \vee \cdots \vee \Phi_n$ $\Phi_i \vee \Phi_j = H(\Phi_i) + H(\Phi_j) - H(\Phi_i)H(\Phi_j)$	$\Phi = \Phi_1 \wedge \Phi_2 \wedge \cdots \wedge \Phi_n$ $\Phi_i \wedge \Phi_j = H(\Phi_i)H(\Phi_j)$	Differentiable everywhere	Not normalized

Another definition of the R_p function is written as

$$
\begin{cases}
\Phi_1 \vee_p \Phi_2 = \Phi_1 + \Phi_2 + \left(\Phi_1^p + \Phi_2^p\right)^{\frac{1}{p}} \\
\Phi_1 \wedge_p \Phi_2 = \Phi_1 + \Phi_2 - \left(\Phi_1^p + \Phi_2^p\right)^{\frac{1}{p}}
\end{cases}
\tag{3.55}
$$

It is also differentiable everywhere except the points $\{x \mid \Phi_1(x) = \Phi_2(x) = 0\}$ and the normalization is improved to order $p - 1$.

When more than two level-set functions are involved, Boolean operations can be handled sequentially or totally to construct the R-function by the so-called n-ary R-conjunction and R-disjunction given below.

$$
\begin{cases}
\bigvee\limits_{i=1}^{n} \Phi_i = \sum\limits_{i=1}^{n} \Phi_i^m (\Phi_i - |\Phi_i|) - \prod\limits_{i=1}^{n} \Phi_i^m (-1)^m (|\Phi_i| - \Phi_i) \\
\bigwedge\limits_{i=1}^{n} \Phi_i = \sum\limits_{i=1}^{n} (-1)^m \Phi_i^m (\Phi_i - |\Phi_i|) + \prod\limits_{i=1}^{n} \Phi_i^m (\Phi_i - |\Phi_i|)
\end{cases}
\tag{3.56}
$$

where m is an integer.

3.3.2 Ricci function

The Ricci function, also called the P-norm function, is another form of functional operation. It is named after the Italian researcher A. Ricci.

In mathematics, the P-norm of a vector $x = (x_1, x_2, \ldots, x_n) \in R^n$ is defined as

$$
\|x\|_p = \left(|x_1|^p + |x_2|^p + \cdots + |x_n|^p\right)^{1/p}
\tag{3.57}
$$

where p is a real number satisfying $p \geq 1$. It is well known that the limit of P-norm for $p \to +\infty$ is the max norm. Due to the fact that the Boolean union and intersection correspond to the max and min functions in mathematics, constructive geometry can be described approximately with P-norm.

Inspired from this, the so-called Ricci function of n level-set functions Φ_1, $\Phi_2, \ldots, \Phi_n$ is written as

$$
\begin{cases}
\Phi_1 \vee_{PN} \Phi_2 \vee_{PN} \cdots \vee_{PN} \Phi_n = \left(\Phi_1^p + \Phi_2^p + \cdots + \Phi_n^p\right)^{1/p} \\
\Phi_1 \wedge_{PN} \Phi_2 \wedge_{PN} \cdots \wedge_{PN} \Phi_n = \left(\Phi_1^{-p} + \Phi_2^{-p} + \cdots + \Phi_n^{-p}\right)^{-1/p}
\end{cases}
\tag{3.58}
$$

In the above equation, it is demanded that the level-set functions Φ_1, $\Phi_2, \ldots, \Phi_n$ be positive everywhere in the concerned domain. However, when all implicit functions $\Phi_1, \Phi_2, \ldots, \Phi_n$ satisfy

$$
\begin{cases}
\Phi_i(x) > 0, & x \in \Omega_i \\
\Phi_i(x) = 0, & x \in \partial\Omega_i \\
\Phi_i(x) < 0, & x \in D \backslash \Omega_i
\end{cases}
\tag{3.59}
$$

the Ricci function in Eq. (3.58) can then be rewritten as an equivalent form via an offset and reverse-offset process

$$\begin{cases} \Phi_1 \vee_{PN} \Phi_2 \vee_{PN} \cdots \vee_{PN} \Phi_n = \left[(\Phi_1 + \overline{\Phi})^p + (\Phi_2 + \overline{\Phi})^p + \cdots + (\Phi_n + \overline{\Phi})^p \right]^{1/p} - \overline{\Phi} \\ \Phi_1 \wedge_{PN} \Phi_2 \wedge_{PN} \cdots \wedge_{PN} \Phi_n = \left[(\Phi_1 + \overline{\Phi})^{-p} + (\Phi_2 + \overline{\Phi})^{-p} + \cdots + (\Phi_n + \overline{\Phi})^{-p} \right]^{-1/p} - \overline{\Phi} \end{cases}$$

(3.60)

where $\overline{\Phi}$ is a relatively large value to ensure that $\Phi_i + \overline{\Phi}$ is greater than zero. For example, we can choose $\overline{\Phi} = |\min(\Phi_1, \Phi_2, ..., \Phi_n)|$.

3.3.3 KS function

The KS function was originally proposed by Kreisselmeier and Steinhauser (1978). It often acts as a numerical fitting technique for the accumulated approximation of target functions (James et al., 2009; Lee et al., 2012). The KS function is expressed as

$$\begin{cases} \Phi_1 \vee_{KS} \Phi_2 \vee_{KS} \cdots \vee_{KS} \Phi_n = \frac{1}{p} \ln \left(e^{p\Phi_1} + e^{p\Phi_2} + \cdots + e^{p\Phi_n} \right) \\ \Phi_1 \wedge_{KS} \Phi_2 \wedge_{KS} \cdots \wedge_{KS} \Phi_n = -\frac{1}{p} \ln \left(e^{-p\Phi_1} + e^{-p\Phi_2} + \cdots + e^{-p\Phi_n} \right) \end{cases}$$

(3.61)

where $p > 0$ is a control factor. It has been proved that the KS function defined in Eq. (3.61) is an enveloping approximation of the max and min functions while the real max and min functions are nondifferentiable at the intersection between two arbitrary level-set functions. Fortunately, the KS function polishes the intersection and improves the nondifferentiability owing to factor p. The bigger the value p, the less polished the intersection and the more accurate the approximation using the KS function.

Practically, the following equivalent form is commonly used to avoid numerical overflow and computational errors (Raspanti et al., 2000)

$$\begin{cases} \Phi_1 \vee_{KS} \Phi_2 \vee_{KS} \cdots \vee_{KS} \Phi_n = \frac{1}{p} \ln \left[e^{p(\Phi_1 - \Phi_{max})} + e^{p(\Phi_2 - \Phi_{max})} + \cdots + e^{p(\Phi_n - \Phi_{max})} \right] + \Phi_{max} \\ \Phi_1 \wedge_{KS} \Phi_2 \wedge_{KS} \cdots \wedge_{KS} \Phi_n = -\frac{1}{p} \ln \left[e^{-p(\Phi_1 - \Phi_{min})} + e^{-p(\Phi_2 - \Phi_{min})} + \cdots + e^{-p(\Phi_n - \Phi_{min})} \right] + \Phi_{min} \end{cases}$$

(3.62)

where $\Phi_{max} = \max(\Phi_1, \Phi_2, ..., \Phi_n)$ and $\Phi_{min} = \min(\Phi_1, \Phi_2, ..., \Phi_n)$.

In the case that the level-set functions $\Phi_1(x), \Phi_2(x), ..., \Phi_n(x)$ are signed distance functions with the basic normalization property

$$\|\nabla \Phi_i\| = 1, \quad i = 1, 2, ..., n$$

(3.63)

it can be demonstrated that the KS function holds the bounded normalization property.

In fact, the following gradient relation can be developed for the KS function defined above

$$\|\nabla \Phi_{KS}\|^2 = \left(\frac{\partial \Phi_{KS}}{\partial x_1}\right)^2 + \left(\frac{\partial \Phi_{KS}}{\partial x_2}\right)^2 + \cdots + \left(\frac{\partial \Phi_{KS}}{\partial x_m}\right)^2$$

$$= \frac{\left(\sum\limits_{i=1}^{n} e^{p\Phi_i}\frac{\partial \Phi_i}{\partial x_1}\right)^2}{\left(\sum\limits_{i=1}^{n} e^{p\Phi_i}\right)^2} + \frac{\left(\sum\limits_{i=1}^{n} e^{p\Phi_i}\frac{\partial \Phi_i}{\partial x_2}\right)^2}{\left(\sum\limits_{i=1}^{n} e^{p\Phi_i}\right)^2} + \cdots + \frac{\left(\sum\limits_{i=1}^{n} e^{p\Phi_i}\frac{\partial \Phi_i}{\partial x_m}\right)^2}{\left(\sum\limits_{i=1}^{n} e^{p\Phi_i}\right)^2}$$

$$= \frac{\sum\limits_{i=1}^{n}\left(e^{p\Phi_i}\right)^2\|\nabla \Phi_i\|^2 + \sum\limits_{i=1}^{n-1}\sum\limits_{j=i+1}^{n}\left(2e^{p\Phi_i}e^{p\Phi_j}\sum\limits_{k=1}^{m}\frac{\partial \Phi_i}{\partial x_k}\frac{\partial \Phi_j}{\partial x_k}\right)}{\left(\sum\limits_{i=1}^{n} e^{p\Phi_i}\right)^2}$$

$$= \frac{\sum\limits_{i=1}^{n}\left(e^{p\Phi_i}\right)^2 + \sum\limits_{i=1}^{n-1}\sum\limits_{j=i+1}^{n}2e^{p\Phi_i}e^{p\Phi_j} + \sum\limits_{i=1}^{n-1}\sum\limits_{j=i+1}^{n}\left(2e^{p\Phi_i}e^{p\Phi_j}\left(\sum\limits_{k=1}^{m}\frac{\partial \Phi_i}{\partial x_k}\frac{\partial \Phi_j}{\partial x_k}-1\right)\right)}{\left(\sum\limits_{i=1}^{m} e^{p\Phi_i}\right)^2}$$

$$= 1 + \frac{\sum\limits_{i=1}^{n-1}\sum\limits_{j=i+1}^{n}\left(2e^{p\Phi_i}e^{p\Phi_j}\left(\sum\limits_{k=1}^{m}\frac{\partial \Phi_i}{\partial x_k}\frac{\partial \Phi_j}{\partial x_k}-1\right)\right)}{\left(\sum\limits_{i=1}^{n} e^{p\Phi_i}\right)^2}$$

$$(3.64)$$

Based on the inequality,

$$\sum\limits_{k=1}^{m}\left(\frac{\partial \Phi_i}{\partial x_k}-\frac{\partial \Phi_j}{\partial x_k}\right)^2 \geq 0, \quad i=1,2,\ldots,n-1; \ j=i+1,\ldots,n \qquad (3.65)$$

the following relation can be derived.

$$\sum\limits_{k=1}^{m}\frac{\partial \Phi_i}{\partial x_k}\frac{\partial \Phi_j}{\partial x_k}-1 \leq 0, \quad i=1,2,\ldots,n-1; \ j=i+1,\ldots,n \qquad (3.66)$$

In view of Eq. (3.64), the module of the gradient of the KS function thus satisfies the bounded normalization property

$$0 \leq \|\nabla \Phi_{KS}\| \leq 1 \qquad (3.67)$$

3.3.4 Step function

Given an implicit function $\Phi(x)$, the approximate step function can be defined as

$$H(\Phi) = \begin{cases} 1, & \Phi \geq \Delta \\ \dfrac{1}{2} + \dfrac{\Phi}{2\Delta}, & -\Delta < \Phi < \Delta \\ 0, & \Phi \leq -\Delta \end{cases} \tag{3.68}$$

where Δ is a very small positive number. When it tends to zero, the above approximate step function will become the exact Heaviside function. Given two implicit functions Φ_1 and Φ_2, the new level-set functions constructed by the Boolean operations of step functions are defined as

$$\begin{cases} \Phi_1 \vee_{ST} \Phi_2 = H(\Phi_1) + H(\Phi_2) - H(\Phi_1)H(\Phi_2) \\ \Phi_1 \wedge_{ST} \Phi_2 = H(\Phi_1)H(\Phi_2) \end{cases} \tag{3.69}$$

This formulation is characterized by the fact that all the values of level-set functions are normalized between 0 and 1 to approximate the exact boundary of the physical domain. The zero level-set will tend to the exact boundary whenever Δ becomes smaller and smaller.

Example 3.8 Boolean intersection operation between $[-4, 4]$ and $[-4, 0]$
Here, R-functions, the Ricci function, the KS function, and the step function are used to obtain the level-set function resulting from the Boolean intersection operation between intervals $[-4, 4]$ and $[-4, 0]$. The intervals can be defined by level-set functions

$$\begin{cases} \Phi_1(x) = 4 - |x| \geq 0 \\ \Phi_2(x) = 2 - |x+2| \geq 0 \end{cases} \tag{3.70}$$

All functions and corresponding one-order derivatives are shown in Fig. 3.11.

Example 3.9 Boolean difference operation between a circle and a supershape
A simple example is shown in Fig. 3.12 to illustrate the effectiveness of R-functions, the Ricci function, the KS function, and the step function in CSG modeling. Suppose two featured primitives related to domains Ω_1 and Ω_2 are implicitly described by Φ_1 and Φ_2

$$\Phi_1 = 1 - x^2 - y^2, \quad \Phi_2 = 1 - x^{1/2} - y^{1/2} \tag{3.71}$$

The new domain Ω constructed by the Boolean difference operation is shown in Fig. 3.12A. The positive parts of the calculated level-set functions are plotted in Fig. 3.12B–G.

3.4   Examples of modeling 2D and 3D mechanical parts

Example 3.10 KS function for modeling 2D mechanical parts
For the purpose of illustration, the hierarchical construction trees of two 2D mechanical parts are given in Figs. 3.13A and 3.14A, where all geometries at the root nodes are represented by the signed distance functions.

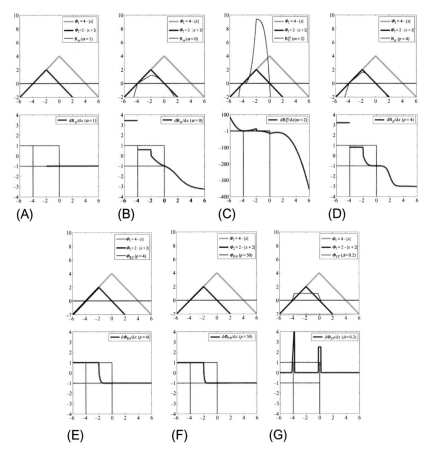

FIG. 3.11 Boolean intersection operation between $[-4, 4]$ and $[-4, 0]$ by means of R-functions, Ricci function, KS function, and step function (upper: level-set function; lower: derivative of level-set function): (A) R_α function ($\alpha = 1$); (B) R_α function ($\alpha = 0$); (C) R_0^m function ($m = 2$); (D) R_p function ($p = 4$); (E) KS function ($p = 4$); (F) Ricci function ($p = 50$); and (G) step function ($\Delta = 0.2$).

The KS function is used to calculate the level-set function of the mechanical parts. The hierarchical KS functions can be constructed as follows.

- 2D mechanical part 1

$$
\begin{cases}
\Phi_{2,1} = \Phi_{1,1} \vee \Phi_{1,2} \vee \cdots \vee \Phi_{1,6} = \dfrac{1}{p} \ln \left(\displaystyle\sum_{i=1}^{6} e^{p \cdot \Phi_{1,i}} \right) \\[2mm]
\Phi_{2,2} = \Phi_{1,7} \vee \Phi_{1,8} \vee \cdots \vee \Phi_{1,13} = \dfrac{1}{p} \ln \left(\displaystyle\sum_{i=7}^{13} e^{p \cdot \Phi_{1,i}} \right) \\[2mm]
\Phi_{3,1} = \Phi_{2,1} \wedge (-\Phi_{2,2}) = -\dfrac{1}{p} \ln \left(e^{-p \cdot \Phi_{2,1}} + e^{-p \cdot (-\Phi_{2,2})} \right)
\end{cases}
\tag{3.72}
$$

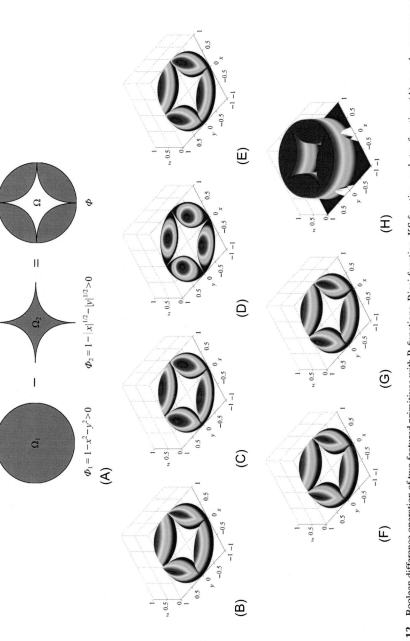

FIG. 3.12 Boolean difference operation of two featured primitives with R-functions, Ricci function, KS function, and step function: (A) complex geometry shape constructed by Boolean difference; (B) R_α function ($\alpha = 1$); (C) R_α function ($\alpha = 0$); (D) R_0^m function ($m = 2$); (E) R_p function ($p = 4$); (F) KS function ($p = 4$); (G) Ricci function ($p = 50$); and (H) step function ($\Delta = 0.2$).

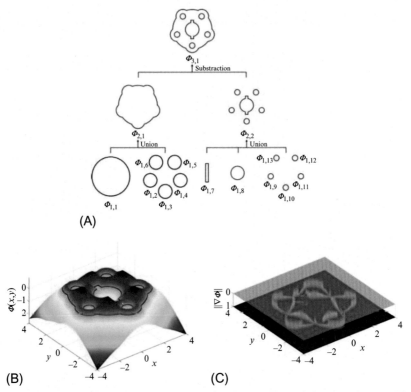

FIG. 3.13 2D mechanical part 1 described by KS function: (A) hierarchical construction tree; (B) higher-dimensional KS function; and (C) gradient module.

- 2D mechanical part 2

$$
\begin{cases}
\Phi_{2,4} = \Phi_{1,1} \vee \Phi_{1,2} \vee \Phi_{1,3} = \dfrac{1}{p} \ln\left(\displaystyle\sum_{i=1}^{3} e^{p \cdot \Phi_{1,i}} \right) \\[2mm]
\Phi_{2,5} = \Phi_{1,4} \vee \Phi_{1,5} \vee \Phi_{1,6} = \dfrac{1}{p} \ln\left(\displaystyle\sum_{i=4}^{6} e^{p \cdot \Phi_{1,i}} \right) \\[2mm]
\Phi_{3,1} = \Phi_{2,1} \wedge (-\Phi_{2,2}) = -\dfrac{1}{p} \ln\left(e^{-p \cdot \Phi_{2,1}} + e^{-p \cdot (-\Phi_{2,2})} \right) \\[2mm]
\Phi_{3,2} = \Phi_{2,4} \wedge (-\Phi_{2,3}) \wedge (-\Phi_{2,5}) = -\dfrac{1}{p} \ln\left(e^{-p \cdot \Phi_{2,4}} + e^{-p \cdot (-\Phi_{2,3})} + e^{-p \cdot (-\Phi_{2,5})} \right) \\[2mm]
\Phi_{4,1} = \Phi_{3,1} \vee \Phi_{3,2} \vee \Phi_{3,3} = \dfrac{1}{p} \ln\left(\displaystyle\sum_{i=1}^{3} e^{p \cdot \Phi_{3,i}} \right)
\end{cases}
$$

$$(3.73)$$

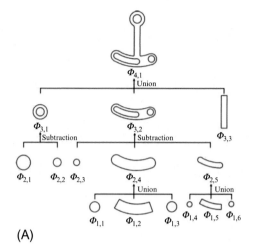

(A)

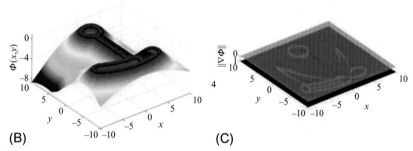

(B)   (C)

FIG. 3.14 2D mechanical part 2 described by KS function: (A) hierarchical construction tree; (B) higher-dimensional KS function; and (C) gradient module.

The constructed KS functions with $p = 200$ are shown in Figs. 3.13B and 3.14B, correspondingly. The bounded normalization property of the gradient module in the interval $[0, 1]$ is verified in Figs. 3.13C and 3.14C, respectively. Moreover, the iso-contours between $[-\Delta, \Delta]$ for the mechanical part 1 at $\Delta = 0.2$ and for the mechanical part 2 at $\Delta = 0.3$ with different values of factor p are given in Figs. 3.15 and 3.16, respectively. Because the signed distance functions are used to represent basic geometries for the hierarchical construction of the KS function, the iso-contours of both mechanical parts are nearly equidistant and tend to be equidistant with the increase of factor p. In contrast, a general level-set function often produces nonequidistant iso-contours.

Example 3.11 Comparison of the R-function and KS function for modeling 3D mechanical parts
Here, the R-function and KS function are also applied for the implicit description of 3D mechanical parts. First, the mechanical parts are decomposed.

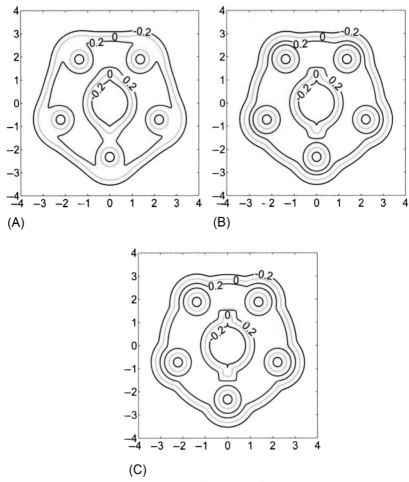

FIG. 3.15 Effects of factor p upon the approximation accuracy and quasiequidistant iso-contours of the KS function for mechanical part 1: (A) $p=5$; (B) $p=10$; and (C) $p=20$.

Corresponding hierarchical constructions are stored in CSG tree structures in Figs. 3.9 and 3.17. The level-set function-based Boolean operations are carried out from the bottom to the top level hierarchically.

The zero iso-surfaces of both the R-function and KS function are shown in Figs. 3.18 and 3.19. Although both functions are capable of describing 3D mechanical parts, differences exist. The R-function is able to accurately describe the geometrical shape while the differentiability is not fully guaranteed at the intersection points of related level-set functions. On the contrary, the KS function is an approximation while the differentiability is guaranteed.

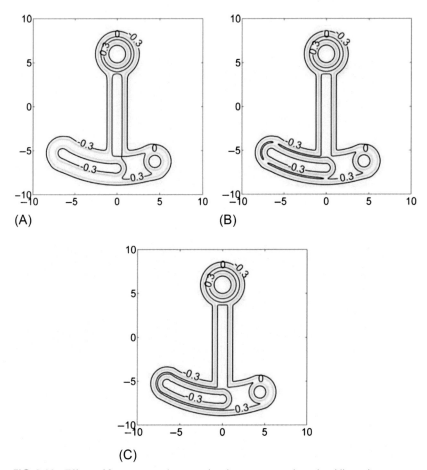

FIG. 3.16 Effects of factor p upon the approximation accuracy and quasiequidistant iso-contours of the KS function for mechanical part 2: (A) $p=5$; (B) $p=10$; and (C) $p=20$.

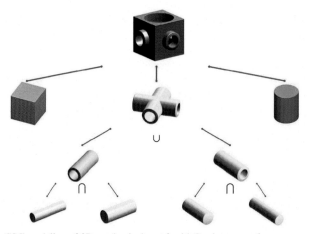

FIG. 3.17 CSG modeling of 3D mechanical part 2 with Boolean operations.

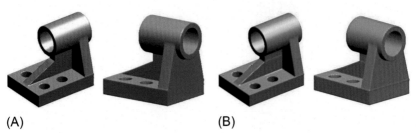

(A)   (B)

FIG. 3.18 3D mechanical part 1 described with R-function and KS function: (A) zero-isosurface of R_0 function; and (B) zero-isosurface of KS function ($p = 50$).

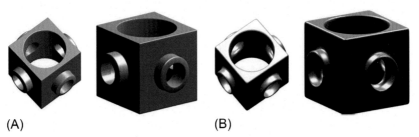

(A)   (B)

FIG. 3.19 3D mechanical part 2 described with R-function and KS function: (A) zero-isosurface of R_0 function; and (B) zero-isosurface of KS function ($p = 50$).

References

James, K.A., Hansen, J.S., Martins, J.R., 2009. Structural topology optimization for multiple load cases using a dynamic aggregation technique. Eng. Optim. 41 (12), 1103–1118.

Kreisselmeier, G., Steinhauser, R., 1978. In flight tests of a parameter insensitive controller. Forschungsber. Dtsch. Forsch. Versuchsanst. LuftRaumfahrt. 7.

Kumar, A.V., Padmanabhan, S., Burla, R., 2008. Implicit boundary method for finite element analysis using non-conforming mesh or grid. Int. J. Numer. Methods Eng. 74 (9), 1421–1447.

Lee, E., James, K.A., Martins, J.R., 2012. Stress-constrained topology optimization with design-dependent loading. Struct. Multidiscip. Optim. 46 (5), 647–661.

Pasko, G.I., Pasko, A.A., Kunii, T.L., 2005. Bounded blending for function-based shape modeling. IEEE Comput. Graph. Appl. 25 (2), 36–45.

Raspanti, C.G., Bandoni, J.A., Biegler, L.T., 2000. New strategies for flexibility analysis and design under uncertainty. Comput. Chem. Eng. 24 (9–10), 2193–2209.

Reiner, T., Mückl, G., Dachsbacher, C., 2011. Interactive modeling of implicit surfaces using a direct visualization approach with signed distance functions. Comput. Graph. 35 (3), 596–603.

Requicha, A.A.G., Voelcker, H.B., 1977. Constructive solid geometry. Tech. Memo. 25. Production Automation Project, Univ. Rochester, New York.

Ricci, A., 1973. A constructive geometry for computer graphics. Comput. J. 16 (2), 157–160.

Rvachev, V.L., 1982. Theory of R-Functions and Some Applications. Naukova Dumka (in Russian).

Shapiro, V., 1991. Theory of R-Functions and Applications: A Primer. Cornell University.

Shapiro, V., 2007. Semi-analytic geometry with R-functions. Acta Numer. 16, 239–303.

Shapiro, V., Tsukanov, I., 1999, June. Implicit functions with guaranteed differential properties. In: Proceedings of the Fifth ACM Symposium on Solid Modeling and Applications, pp. 258–269.

Chapter 4

B-spline finite cell method for structural analysis

4.1 Introduction to B-spline finite cell method

The finite cell method (FCM) was initially developed by Parvizian et al. (2007) and Düster et al. (2008) for structural analysis with fixed mesh. It is based on the concept of fictitious domain that extends the concerned physical fields from an arbitrarily shaped physical domain to a regular embedding domain. The difference between FCM and traditional fictitious domain methods (Del Pino and Pironneau, 2003; Ramiere et al., 2007) is that the FCM uses high-order shape functions like the Legendre polynomial to approximate the extended physical fields. Specially, the B-spline FCM means that B-spline basis functions are used to interpolate the physical field. B-spline basis functions are adopted because their high-order and high-continuity properties that can ensure an increased per-degree-of-freedom accuracy in many cases (Ruess et al., 2013). Comparatively, the simplified version of XFEM (Daux et al., 2000; Wei et al., 2010; Guo et al., 2011; Zhang et al., 2013; Wang and Li, 2013) can be regarded as a variant of FCM with shape functions of order one.

4.1.1 B-spline basis function

Consider the bivariate B-spline basis functions for 2D problems. The expression corresponds to

$$M_k(\xi, \eta) = B_{i,p}(\xi) \cdot B_{j,q}(\eta) \quad k = i + n(j - 1) \tag{4.1}$$

where $B_{i,p}(\xi)$ and $B_{j,q}(\eta)$ are two univariate B-spline basis functions defined in the parametric coordinate system over knot vectors $\Xi = \{\xi_1, \xi_2, \ldots, \xi_{n+p+1}\}$ and $H = \{\eta_1, \eta_2, \ldots, \eta_{m+q+1}\}$, respectively. The expression of the univariate B-spline basis function $B_{i,p}(\xi)$ and $B_{j,q}(\eta)$ holds the well-known recurrence relation in Eq. (2.55). p and q denote the polynomial orders. n and m are numbers of $B_{i,p}(\xi)$ and $B_{j,q}(\eta)$, respectively. The support of the univariate B-spline basis function $B_{i,p}(\xi)$ reaches between $\xi_i \leq \xi \leq \xi_{i+p+1}$. In addition, $B_{i,p}(\xi)$ is $(p-1)$-times continuously differentiable on its support.

The Feature-driven Method for Structural Optimization. https://doi.org/10.1016/B978-0-12-821330-8.00004-3

4.1.2 Basic theory of B-spline finite cell method

Without loss of generality, consider a 2D linear thermoelastic structure defined in the physical domain Ω with its Lipschitz continuous boundary $\partial\Omega$, as illustrated in Fig. 4.1. Assume that the thermal and elastic quantities are marked with the superscripts "th" and "el," respectively. The temperature distribution corresponds to the strong form of the boundary value problem

$$\begin{cases} -\nabla \cdot (\kappa \nabla T) = s & \text{in } \Omega \\ T = r & \text{on } \Gamma_D^{\text{th}} \\ \nabla T \cdot \boldsymbol{n} = q & \text{on } \Gamma_N^{\text{th}} \end{cases} \quad (4.2)$$

where κ, s, T, r, $\boldsymbol{n}$, and q denote the thermal heat conduction coefficient, the heat source, the unknown temperature, the prescribed temperature on Γ_D^{th}, the unit outward normal vector on Γ_N^{th}, and the prescribed normal heat flux on Γ_N^{th}.

Besides, elastic responses are governed by the strong form of the boundary value problem

$$\begin{cases} -\nabla \cdot \boldsymbol{\sigma} = \boldsymbol{f} & \text{in } \Omega \\ \boldsymbol{u} = \boldsymbol{g} & \text{on } \Gamma_D^{\text{el}} \\ \boldsymbol{\sigma} \cdot \boldsymbol{n} = \boldsymbol{\tau} & \text{on } \Gamma_N^{\text{el}} \\ \boldsymbol{\varepsilon} = (\nabla \boldsymbol{u} + \nabla^T \boldsymbol{u})/2 & \text{in } \Omega \\ \boldsymbol{\varepsilon}^{\text{th}} = \gamma \cdot (T - T_0) \cdot \boldsymbol{I} & \text{in } \Omega \\ \boldsymbol{\sigma} = \boldsymbol{C} : \left(\boldsymbol{\varepsilon} - \boldsymbol{\varepsilon}^{\text{th}} \right) & \text{in } \Omega \end{cases} \quad (4.3)$$

where $\boldsymbol{\sigma}$, $\boldsymbol{f}$, and $\boldsymbol{u}$ denote the stress tensor, body force in Ω, and displacement. $\boldsymbol{g}$, $\boldsymbol{n}$, and $\boldsymbol{\tau}$ denote the prescribed displacement on Γ_D^{el}, the unit outward normal vector on Γ_N^{el}, and the prescribed boundary tractions on Γ_N^{el}. $\boldsymbol{\varepsilon}$, $\boldsymbol{\varepsilon}^{\text{th}}$, γ, T_0, and $\boldsymbol{C}$ denote the mechanical strain tensor, the thermal strain tensor, the thermal

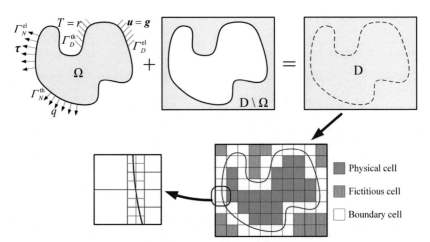

FIG. 4.1 Illustration of a 2D thermoelastic structure Ω embedded in the fictitious domain D with regular cell discretization.

expansion coefficient, the stress-free reference temperature, and the elastic constitutive tensor. I is the identity matrix implying that the temperature distribution never influences the shear strain.

According to the fictitious domain concept, the physical domain Ω is extended into an embedding domain D of simple rectangular form. The latter can be easily discretized by structured or Cartesian meshes and then used as the computing model. The structured mesh is called cells to distinguish with elements in FEM. Three types of cells exist according to their positions relative to the physical boundary $\partial\Omega$, that is, physical cells located inside $\partial\Omega$, fictitious cells located outside $\partial\Omega$, and boundary cells cut by $\partial\Omega$.

The discretization of the embedding domain is a mapping of the standard parametric domain $\xi \times \eta = [0, 1] \times [0, 1]$, as shown in Fig. 4.2. Any arbitrary point $P = (x, y)^T \in D$ can be interpolated by the following mapping relation

$$P = \sum_{k=1}^{m \times n} M_k(\xi, \eta)P_k \tag{4.4}$$

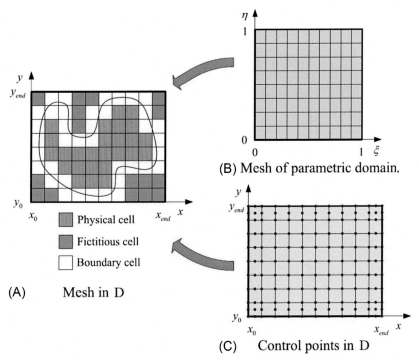

FIG. 4.2 Mapping from standard parametric domain $\xi \times \eta$ to embedding domain D: (A) regular cell discretization of embedding domain D; (B) parametric mesh used to define bivariate B-spline basis functions $M_k(\xi,\eta)$; and (C) mesh of control points P_k with polynomial orders $p = q = 2$, and uniform and open knot vectors $\varXi$ and H.

with $P_k=(x_k,y_k)^T$ denoting the kth control point shown in Fig. 4.2C. Furthermore, Eq. (4.4) can be simplified as a linear mapping relation for the determination of points $P \in D$

$$P=S(\xi,\eta)=\begin{bmatrix} x_0 + \xi(x_{end}-x_0) \\ y_0 + \eta(y_{end}-y_0) \end{bmatrix} \tag{4.5}$$

where x_0, x_{end}, y_0, and y_{end} define four extreme points of domain D shown in Fig. 4.2.

Weak forms of Eqs. (4.2), (4.3) in the embedding domain D can then be written as

$$a^{\text{th}}(T,R)=l^{\text{th}}(R) \tag{4.6}$$

with

$$\begin{cases} a^{\text{th}}(T,R)=\int_D \nabla R \cdot \alpha\kappa \cdot \nabla T d\Omega \\ l^{\text{th}}(R)=\int_D R \cdot \alpha s d\Omega + \int_{\Gamma_N^{\text{th}}} R \cdot q d\Gamma \end{cases} \tag{4.7}$$

and

$$a^{\text{el}}(\boldsymbol{u},\boldsymbol{v})=l^{\text{el}}(\boldsymbol{v},T) \tag{4.8}$$

with

$$\begin{cases} a^{\text{el}}(\boldsymbol{u},\boldsymbol{v})=\int_D \boldsymbol{\varepsilon}(\boldsymbol{v}):\alpha\boldsymbol{C}:\boldsymbol{\varepsilon}(\boldsymbol{u})d\Omega \\ l^{\text{el}}(\boldsymbol{v},T)=\int_D \boldsymbol{v}\cdot\alpha\boldsymbol{f}d\Omega + \int_{\Gamma_N^{\text{el}}} \boldsymbol{v}\cdot\boldsymbol{\tau}d\Gamma + \int_D \boldsymbol{\varepsilon}(\boldsymbol{v}):\alpha\boldsymbol{C}:\boldsymbol{\varepsilon}^{\text{th}}(T)d\Omega \end{cases} \tag{4.9}$$

In the above expressions, the scalar α acts as a Heaviside function and is defined as

$$\alpha=\begin{cases} 1 & \text{in } \Omega \\ 0 & \text{in } D\backslash\Omega \end{cases} \tag{4.10}$$

Admissible trial functions for scalar temperature R and vectoral displacement $\boldsymbol{v}$ are defined in the one-order Sobolev space $H^1(D)$ and $\boldsymbol{H}^1(D)$, respectively.

$$\begin{cases} S_T=\{T|\, T\in H^1(D),\ T=r \text{ on } \Gamma_D^{\text{th}}\} \\ S_R=\{R|\, R\in H^1(D),\ R=0 \text{ on } \Gamma_D^{\text{th}}\} \end{cases} \tag{4.11}$$

$$\begin{cases} S_u=\{\boldsymbol{u}|\, \boldsymbol{u}\in\boldsymbol{H}^1(D),\ \boldsymbol{u}=\boldsymbol{g} \text{ on } \Gamma_D^{\text{el}}\} \\ S_v=\{\boldsymbol{v}|\, \boldsymbol{v}\in\boldsymbol{H}^1(D),\ \boldsymbol{v}=\boldsymbol{0} \text{ on } \Gamma_D^{\text{el}}\} \end{cases} \tag{4.12}$$

Suppose S_T^h, S_R^h, S_u^h, and S_v^h are finite-dimensional subspaces of spaces S_T, S_R, S_u, and S_v, respectively. The discrete problems of Eqs. (4.6), (4.8) are then to find discrete solutions $T^h \in S_T^h$ and $u^h \in S_u^h$ that satisfy

$$\forall R^h \in S_R^h, \quad a^{th}\left(T^h, R^h\right) = l^{th}\left(R^h\right) \tag{4.13}$$

and

$$\forall v^h \in S_v^h, \quad a^{el}\left(u^h, v^h\right) = l^{el}\left(v^h, T^h\right) \tag{4.14}$$

Regarding the temperature field, when Γ_D^{th} belongs to a portion of the embedding domain's boundary ∂D, R^h can be interpolated as the test temperature by using those basis functions M^{th} that vanish at Γ_D^{th}. Otherwise, M^{th} should be revised by employing the so-called weighted extended B-spline method (Web method), which will be addressed in detail in Section 4.2. Without loss of generality, assume M^{th} are constructed such that $M^{th}|_{\Gamma_D^{th}} = 0$. There exists an arbitrary vector A such that

$$R^h = M^{th}A \tag{4.15}$$

where M^{th} refers to the shape function matrix at $m \times n$ discrete control points in Fig. 4.2. In the B-spline FCM, the B-spline basis functions in Eq. (4.1) are adopted to construct M^{th} as

$$M^{th} = [M_1\ M_2\ \cdots\ M_{m \times n}] \tag{4.16}$$

Assume T denotes the vector of unknown coefficients, that is, temperatures at discrete control points. For a given function r^h satisfying $r^h|_{\Gamma_D^{th}} = r$, the approximated temperature can be expressed as

$$T^h = M^{th}T + r^h \tag{4.17}$$

For a 2D problem, the gradient of T^h needed in the discrete form of Eq. (4.7) can be calculated by

$$\nabla T^h = \left\{ \begin{array}{c} \dfrac{\partial T^h}{\partial x} \\[2mm] \dfrac{\partial T^h}{\partial y} \end{array} \right\} = B^{th}T + \nabla r^h \tag{4.18}$$

where B^{th} and ∇r^h are expressed as

$$B^{th} = \begin{bmatrix} \dfrac{\partial}{\partial x} \\[2mm] \dfrac{\partial}{\partial y} \end{bmatrix} M^{th} = \begin{bmatrix} \dfrac{\partial M_1}{\partial x} & \dfrac{\partial M_2}{\partial x} & \cdots & \dfrac{\partial M_{m \times n}}{\partial x} \\[3mm] \dfrac{\partial M_1}{\partial y} & \dfrac{\partial M_2}{\partial y} & \cdots & \dfrac{\partial M_{m \times n}}{\partial y} \end{bmatrix} \tag{4.19}$$

$$\nabla r^h = \begin{bmatrix} \dfrac{\partial r^h}{\partial x}, & \dfrac{\partial r^h}{\partial y} \end{bmatrix}^T \tag{4.20}$$

Based on the Ritz-Galerkin method, the thermal governing system of Eq. (4.6) corresponds to the following discretization form

$$\boldsymbol{K}^{\text{th}}\boldsymbol{T} = \boldsymbol{F}^{\text{th}} \tag{4.21}$$

where $\boldsymbol{K}^{\text{th}}$ and $\boldsymbol{F}^{\text{th}}$ are the global conductivity matrix and the global heat flux vector

$$\begin{cases} \boldsymbol{K}^{\text{th}} = \displaystyle\int_{\text{D}} \left(\boldsymbol{B}^{\text{th}}\right)^T \alpha\kappa\boldsymbol{B}^{\text{th}}\mathrm{d}\Omega \\ \boldsymbol{F}^{\text{th}} = \displaystyle\int_{\text{D}} \left(\boldsymbol{M}^{\text{th}}\right)^T \alpha s\mathrm{d}\Omega + \int_{\Gamma_N^{\text{th}}} \left(\boldsymbol{M}^{\text{th}}\right)^T q\mathrm{d}\Gamma - \int_{\text{D}} \left(\boldsymbol{B}^{\text{th}}\right)^T \alpha\kappa\nabla r^h\mathrm{d}\Omega \end{cases} \tag{4.22}$$

Similarly, the displacement field related to Eq. (4.8) is written as

$$\boldsymbol{u}^{\text{h}} = \boldsymbol{M}^{\text{el}}\boldsymbol{U} + \boldsymbol{g}^{\text{h}} \tag{4.23}$$

where $\boldsymbol{U}$ denotes the displacement vector at discretized control points. $\boldsymbol{g}^{\text{h}}$ refers to the vector of the boundary value function satisfying $\boldsymbol{g}^{\text{h}}|_{\Gamma_B^{\text{el}}} = \boldsymbol{g}$. The shape function matrix $\boldsymbol{M}^{\text{el}}$ interpolating the displacement field is expressed as

$$\boldsymbol{M}^{\text{el}} = \begin{bmatrix} M_1 & 0 & \cdots & M_{m\times n} & 0 \\ 0 & M_1 & \cdots & 0 & M_{m\times n} \end{bmatrix} \tag{4.24}$$

The strain vector is then calculated by

$$\boldsymbol{\varepsilon}\left(\boldsymbol{u}^{\text{h}}\right) = \begin{Bmatrix} \varepsilon_x \\ \varepsilon_y \\ \gamma_{xy} \end{Bmatrix} = \begin{Bmatrix} \dfrac{\partial u_x^{\text{h}}}{\partial x} \\[2mm] \dfrac{\partial u_y^{\text{h}}}{\partial y} \\[2mm] \dfrac{\partial u_x^{\text{h}}}{\partial y} + \dfrac{\partial u_y^{\text{h}}}{\partial x} \end{Bmatrix} = \boldsymbol{B}^{\text{el}}\boldsymbol{U} + \partial\boldsymbol{g}^{\text{h}} \tag{4.25}$$

where u_x^{h} and u_y^{h} are the 2D components of $\boldsymbol{u}^{\text{h}}$ while the strain-displacement matrix $\boldsymbol{B}^{\text{el}}$ and the last term $\partial\boldsymbol{g}^{\text{h}}$ correspond to

$$\boldsymbol{B}^{\text{el}} = \begin{bmatrix} \dfrac{\partial}{\partial x} & 0 \\[2mm] 0 & \dfrac{\partial}{\partial y} \\[2mm] \dfrac{\partial}{\partial y} & \dfrac{\partial}{\partial x} \end{bmatrix} \quad \boldsymbol{M}^{\text{el}} = \begin{bmatrix} \dfrac{\partial M_1}{\partial x} & 0 & \cdots & \dfrac{\partial M_{m\times n}}{\partial x} & 0 \\[2mm] 0 & \dfrac{\partial M_1}{\partial y} & \cdots & 0 & \dfrac{\partial M_{m\times n}}{\partial y} \\[2mm] \dfrac{\partial M_1}{\partial y} & \dfrac{\partial M_1}{\partial x} & \cdots & \dfrac{\partial M_{m\times n}}{\partial y} & \dfrac{\partial M_{m\times n}}{\partial x} \end{bmatrix} \tag{4.26}$$

$$\partial\boldsymbol{g}^{\text{h}} = \begin{bmatrix} \dfrac{\partial g_x^{\text{h}}}{\partial x}, & \dfrac{\partial g_y^{\text{h}}}{\partial y}, & \dfrac{\partial g_x^{\text{h}}}{\partial y} + \dfrac{\partial g_y^{\text{h}}}{\partial x} \end{bmatrix}^T \tag{4.27}$$

Correspondingly, the FCM system of equations can be derived as

$$\boldsymbol{K}^{\mathrm{el}}\boldsymbol{U} = \boldsymbol{F}^{\mathrm{el}} \tag{4.28}$$

where $\boldsymbol{K}^{\mathrm{el}}$ and $\boldsymbol{F}^{\mathrm{el}}$ denote the global stiffness matrix and the global load vector, respectively.

$$\begin{cases} \boldsymbol{K}^{\mathrm{el}} = \displaystyle\int_{\mathrm{D}} \left(\boldsymbol{B}^{\mathrm{el}}\right)^{T} \alpha \boldsymbol{C} \boldsymbol{B}^{\mathrm{el}} \mathrm{d}\Omega \\[2mm] \boldsymbol{F}^{\mathrm{el}} = \displaystyle\int_{\mathrm{D}} \left(\boldsymbol{M}^{\mathrm{el}}\right)^{T} \alpha \boldsymbol{f} \mathrm{d}\Omega + \int_{\Gamma_{N}^{\mathrm{el}}} \left(\boldsymbol{M}^{\mathrm{el}}\right)^{T} \boldsymbol{\tau} \mathrm{d}\Gamma + \int_{\mathrm{D}} \left(\boldsymbol{B}^{\mathrm{el}}\right)^{T} \alpha \boldsymbol{C} \left(\boldsymbol{\varepsilon}^{\mathrm{th}} - \partial \boldsymbol{g}^{\mathrm{h}}\right) \mathrm{d}\Omega \end{cases} \tag{4.29}$$

The $\boldsymbol{K}^{\mathrm{th}}$, $\boldsymbol{F}^{\mathrm{th}}$, $\boldsymbol{K}^{\mathrm{el}}$, and $\boldsymbol{F}^{\mathrm{el}}$ involved in Eqs. (4.22), (4.29) are practically calculated by the Gaussian quadrature over all cells. For physical cells, one such calculation is done in the usual way, but the computing efficiency can be ameliorated by means of the periodic property of shape function. As to fictitious cells, cell conductivity and stiffness matrices are not calculated. Boundary cells should be treated specially by means of the quadtree/octree scheme. To clarify the idea, a boundary cell hierarchically partitioned into subcells with a maximum level of three is illustrated in Fig. 4.1. Consequently, the number of Gauss points is adaptively increased around $\partial \Omega$ along with the subcell refinement to ensure the computing accuracy while the same basis functions are used. More details about the quadtree/octree refinement are given in Section 4.1.3.

4.1.3 Cell refinement with quadtree/octree scheme

The main interest of FCM is the use of a regular and fixed mesh through the whole optimization process, whatever the shape and topology changes of the physical domain Ω are. This is the main difference between the FCM and FEM. As the latter is based on the so-called body-fitted mesh conformal to the physical boundary of a structure, sophisticated remeshing is therefore unavoidable to follow the boundary variation of a structure during the optimization process. This is, in fact, a long-term challenging issue in boundary-based shape and topology optimization.

The FCM grid is fixed and independent of the structural boundary modification. However, Gauss integration points should be enriched for boundary cells to compute their stiffness matrices and conductivity matrices. This is realized by refining the boundary cells using the quadtree (2D problems) or octree (3D problems) scheme (Parvizian et al., 2007; Düster et al., 2008; Schillinger et al., 2012a,b) to identify the involved solid and void portions. This kind of integration scheme thus ensures the accuracy of structural analysis without introducing additional degrees of freedoms (d.o.fs).

Technically, quadtree or octree is a kind of tree data structure method used to partition a domain recursively into four (2D) or eight (3D) children. With the aid of the level-set representation $\Phi(\boldsymbol{x})$ of the concerned boundary, the cell refinement is made easily processed with a simple inside/outside test of cell position. To be specific, as mentioned in Eq. (2.12), the position of an arbitrary node $x \in D$ with respect to the boundary of physical domain can simply be determined according to the sign of the level-set function $\Phi(x)$. Notice that the level-set function can be constructed in different ways to represent a complicated boundary with the aid of Boolean operations widely used in solid modeling. As discussed in Chapter 3, the R-function, KS function, Ricci function, and step function can be used.

As shown in Fig. 4.3, the refinement is made recursively to track the boundary cutting. With the increasingly hierarchical refinement (Schillinger et al., 2012b), boundary cells are partitioned into smaller subcells to improve local stress computing accuracy. Notice that a large number of refinement levels will considerably increase the computing cost while a small number of refinement levels might not ensure the computing accuracy. Thus, the number of refinement levels should be selected properly.

Fig. 4.4 illustrates a boundary cell with 3×3 Gauss integration points in each subcell. Generally, only interior Gauss integration points with $\alpha = 1$ in Eqs. (4.22), (4.29) are considered in the formulation of the stiffness matrix and conductivity matrix. In order to overcome ill-conditioning problems, $\alpha = 0$ is often replaced with a very small positive number $\alpha - 10^{-3} - 10^{-12}$. In comparison with the conventional FEM, a body-fitted mesh is no longer required to follow the modification of the physical boundary during the optimization process. This is indeed the great advantage of the FCM for its design efficiency and versatility.

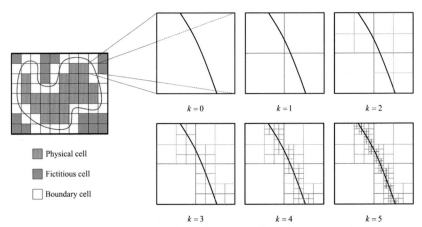

FIG. 4.3 Integration scheme with quadtree refinement of boundary cells (k is the number of refinement levels).

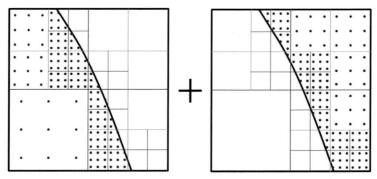

FIG. 4.4 Gauss integration points inside *(left)* and outside *(right)* the physical domain Ω in a boundary cell.

4.2 Imposition of the Dirichlet boundary condition with the Web method

4.2.1 Imposition methods of the Dirichlet boundary condition

As is known, the Dirichlet boundary conditions (DBCs) generally hold two forms, the homogeneous Dirichlet boundary conditions (HDBCs) and the inhomogeneous Dirichlet boundary conditions (IDBCs). The former can be considered as a special case of the latter with zero imposed value. How to properly impose boundary conditions is essential in varieties of numerical analysis methods, such as FEM (Zienkiewicz et al., 1977), XFEM (Moës et al., 2006), the meshfree method (Fernández-Méndez and Huerta, 2004), and FCM (Parvizian et al., 2007; Düster et al., 2008; Schillinger et al., 2012a,b; Zander et al., 2012; Abedian et al., 2013; Joulaian and Düster, 2013; Zhang et al., 2015). This constitutes a key step for the achievement of high solution accuracy. For a conformal FE mesh, one can simply restrain d.o.fs of corresponding nodes on the Dirichlet boundary or resort to the so-called multipoint constraint (Ainsworth, 2001). However, the mesh should be updated to maintain the body-fitted property when the boundary shape changes, as in the case of shape optimization. In the worst case of mesh updating, the mesh distortion and the ill-conditioned element conductivity/stiffness matrix might break down the optimization process. More importantly, boundary conditions are imposed to FE nodes on the concerned boundary in a discrete way and are therefore mesh-dependent.

The FCM has the advantage over the FEM because the former concerns a nonconformal fixed mesh and a high-order shape function, whatever the boundary shape is. However, a direct imposition of DBCs then implies restraining multiple neighboring control points and thus becomes difficult to realize in practice. This is the typical issue to be investigated in the FCM where interpolation functions are defined by high-order shape functions without holding the Kronecker delta property.

Generally speaking, two kinds of approaches have been developed until now. The first one concerns weak imposition methods, including the Lagrange multiplier (Glowinski et al., 1999; Moës et al., 2006; Glowinski and Kuznetsov, 2007), the penalty method (Fernández-Méndez and Huerta, 2004), and Nitsche's method (Nitsche, 1971; Dupire et al., 2010; Embar et al., 2010; Boufflet et al., 2012), which introduced IDBCs indirectly into the Galerkin weak form. However, the Lagrange multiplier might increase the number of variables and the computational complexity. The computing accuracy of the penalty method strongly depends upon the attributed values of the involved penalty parameters. Although large values of penalty parameters theoretically improve the accuracy of imposing DBCs, they have a tendency to yield an ill-conditioned system of equations. Nitsche's method, sometimes referred to as the variationally consistent penalty method (Embar et al., 2010), is preferably considered as a stabilized method. Although the modification of the weak form does not suffer from ill-conditioning and holds the problem scale and symmetry of the system of equations, it is different for each specific problem (Fernández-Méndez and Huerta, 2004). In the work of Zander et al. (2012), Nitsche's method was combined with the FCM to impose DBCs for thermoelastic problems. In the work of Ruess et al. (2013), this method was also combined with FCM and isogeometric analysis using high-order B-spline basis functions.

The second kind of approach consists of directly modifying the interpolation functions to impose DBCs. This is done by introducing an implicit function related to the Dirichlet boundary as the weighting function in the interpolation scheme of the physical field. Comparatively, this approach is more straightforward than the weak imposition method because the weighting function has the effect of fixing the Dirichlet boundary in advance. There are two basic schemes in this kind of approach, that is, the weighted extended B-spline method (Web method) (Höllig et al., 2001, 2005; Höllig and Reif, 2003) and the implicit boundary method (Burla and Kumar, 2008; Kumar et al., 2008).

In the Web method, the weighting function is defined by the R-function to impose HDBCs and the interpolation function itself is originally defined as a linear combination of the basis of B-splines to stabilize the convergence procedure. Nevertheless, this scheme is limited to the imposition of HDBCs. By contrast, the implicit boundary method is to define the weighting function with the step function. Because IDBCs are imposed through interpolation by segmenting the Dirichlet boundary within the discretized mesh, the approximation errors of IDBCs will produce inevitably.

To highlight the existing methods, the imposition of DBCs is summarized in Table 4.1 for Poisson's equation

$$\begin{cases} -\nabla^2 u = f & \text{in } \Omega \\ u = g & \text{on } \Gamma_D \\ \nabla u \cdot \boldsymbol{n} = \tau & \text{on } \Gamma_N \end{cases} \qquad (4.30)$$

TABLE 4.1 Summary of different kinds of methods for imposing DBCs.

Type	Method	Energy functional	Interpolation form	Linear system	Properties
Weak imposition method	Lagrange multiplier (Fernández-Méndez and Huerta 2004)	$\Pi(u) + \int_{\Gamma_D}\lambda(u-g)d\Gamma$	$u^h = NU$	$\begin{bmatrix} K & K_L^T \\ K_L & 0 \end{bmatrix}\begin{bmatrix} U \\ \Lambda \end{bmatrix} = \begin{bmatrix} F \\ F_L \end{bmatrix}$	Additional d.o.fs; Stiffness matrix no longer positive definite
	Penalty method (Fernández-Méndez and Huerta 2004)	$\Pi(u) + \frac{1}{2}\beta\int_{\Gamma_D}(u-g)^2 d\Gamma$	$u^h = NU$	$(K + \beta K_P)U = F + \beta F_P$	Ill-conditioned stiffness matrix for fine meshes; β-dependent accuracy
	Nitsche's method (Fernández-Méndez and Huerta 2004)	$\Pi(u) - \int_{\Gamma_D}\nabla u\cdot n(u-g)d\Gamma + \frac{1}{2}\beta\int_{\Gamma_D}(u-g)^2 d\Gamma$	$u^h = NU$	$(K - K_M - K_M^T + \beta K_P)U = F - F_M + \beta F_P$	Consistent and stabilized weak form; Accuracy depending on β
Weighted interpolation method	Web-method (Höllig et al. 2005)	$\Pi(u)$	$u^h = wNU$	$\bar{K}U = \bar{F}$	Limited to the imposition of HDBCs
	Implicit boundary method (Kumar et al. 2008)	$\Pi(u)$	$u^h = wNU + N\bar{g}$	$\bar{K}U = \bar{F} - F_I$	Approximate imposition of IDBCs
	Weighted FCM method	$\Pi(u)$	$u^h = wNU + G$	$\bar{K}U = \bar{F} - F_W$	Exact imposition of IDBCs
Notations	$\omega = 0$ on Γ_D	$\Pi(u) = \frac{1}{2}\int_\Omega \nabla u\cdot\nabla u\, d\Omega - \int_\Omega ufd\Omega - \int_{\Gamma_N} utd\Gamma$ $K_L = K_P = \int_{\Gamma_D} N^T N d\Gamma$ $F_L = F_P = \int_{\Gamma_D} N^T g d\Gamma$ $\bar{F} = \int_\Omega \bar{N}^T fd\Omega + \int_{\Gamma_N}\bar{N}^T td\Gamma$ $F_W = \int_\Omega \nabla\bar{N}^T\nabla G d\Omega$			$K_M = \int_{\Gamma_D}\nabla N^T n N d\Gamma$ $F_M = \int_{\Gamma_D}\nabla N^T n g d\Gamma$

$\bar{N} = \omega N$

$K = \int_\Omega \nabla N^T\nabla N d\Omega$

$F = \int_\Omega N^T fd\Omega + \int_{\Gamma_N} N^T td\Gamma$

$\bar{K} = \int_\Omega \nabla\bar{N}^T\nabla N d\Omega$

$F_I = \int_\Omega \nabla\bar{N}^T\nabla g d\Omega$

4.2.2 Weighted B-spline finite cell method

In this section, a weighted B-spline FCM is introduced to impose the DBCs exactly by modifying the interpolation functions. The basis functions of the weighted B-spline FCM can be obtained by local and slight modifications of the B-spline basis functions of Eq. (4.1) to impose DBCs exactly.

$$\overline{M}_k^{\text{th}}(\xi, \eta) = \omega^{\text{th}}(x, y)M_k(\xi, \eta) = \omega^{\text{th}}(x, y)B_{i,p}(\xi)B_{j,q}(\eta) \quad k = i + n(j-1) \quad (4.31)$$

$$\overline{M}_k^{\text{el}}(\xi, \eta) = \omega^{\text{el}}(x, y)M_k(\xi, \eta) = \omega^{\text{el}}(x, y)B_{i,p}(\xi)B_{j,q}(\eta) \quad k = i + n(j-1) \quad (4.32)$$

with the weighting function $\omega^{\text{th}}(x,y)|_{\Gamma_D\text{th}} = 0$ and $\omega^{\text{el}}(x,y)|_{\Gamma_D\text{el}} = 0$. Practically, $\omega^{\text{th}}(x, y)$ and $\omega^{\text{el}}(x, y)$ can easily be obtained by revising the level-set functions representing the boundary $\partial\Omega$. The weighted B-spline is a simple form of the Web method studied in Höllig et al. (2001), Höllig and Reif (2003), Höllig et al. (2005). One can refer to Höllig et al. (2001), Höllig and Reif (2003), Höllig et al. (2005) for the advanced version of the Web method written as

$$\overline{M}_k(\xi, \eta) = \omega(x, y)\left(M_k(\xi, \eta) + \sum_{l \in L} e_{k,l}M_l(\xi, \eta)\right) \quad (4.33)$$

in which $e_{k,l}$ acts as extension coefficients to stabilize $\overline{M}_k(\xi, \eta)$. $M_l(\xi, \eta)$ belongs to the neighboring basis functions of $M_k(\xi, \eta)$.

The weighted B-spline basis function results in the following interpolation fields

$$T^{\text{h}}(\boldsymbol{x}) = \omega^{\text{th}}(\boldsymbol{x})\boldsymbol{M}\boldsymbol{T} + r^{\text{h}} = \overline{\boldsymbol{M}}^{\text{th}}\boldsymbol{T} + r^{\text{h}} \quad (4.34)$$

$$\boldsymbol{u}^{\text{h}}(\boldsymbol{x}) = \omega^{\text{el}}(\boldsymbol{x})\boldsymbol{M}\boldsymbol{U} + \boldsymbol{g}^{\text{h}} = \overline{\boldsymbol{M}}^{\text{el}}\boldsymbol{U} + \boldsymbol{g}^{\text{h}} \quad (4.35)$$

Correspondingly, $\boldsymbol{B}^{\text{th}}$ and $\boldsymbol{B}^{\text{el}}$ in Eqs. (4.19), (4.26) correspond to

$$\overline{\boldsymbol{B}}^{\text{th}} = \begin{bmatrix} \dfrac{\partial}{\partial x} \\[2mm] \dfrac{\partial}{\partial y} \end{bmatrix} \overline{\boldsymbol{M}}^{\text{th}}$$

$$= \begin{bmatrix} \dfrac{\partial\omega^{\text{th}}}{\partial x}M_1 + \omega^{\text{th}}\dfrac{\partial M_1}{\partial x} & \dfrac{\partial\omega^{\text{th}}}{\partial x}M_2 + \omega^{\text{th}}\dfrac{\partial M_2}{\partial x} & \cdots & \dfrac{\partial\omega^{\text{th}}}{\partial x}M_{n\times m} + \omega^{\text{th}}\dfrac{\partial M_{n\times m}}{\partial x} \\[4mm] \dfrac{\partial\omega^{\text{th}}}{\partial y}M_1 + \omega^{\text{th}}\dfrac{\partial M_1}{\partial y} & \dfrac{\partial\omega^{\text{th}}}{\partial y}M_2 + \omega^{\text{th}}\dfrac{\partial M_2}{\partial y} & \cdots & \dfrac{\partial\omega^{\text{th}}}{\partial y}M_{n\times m} + \omega^{\text{th}}\dfrac{\partial M_{n\times m}}{\partial y} \end{bmatrix}$$

$$(4.36)$$

$$\overline{\boldsymbol{B}}^{el} = \begin{bmatrix} \dfrac{\partial}{\partial x} & 0 \\[2mm] 0 & \dfrac{\partial}{\partial y} \\[2mm] \dfrac{\partial}{\partial y} & \dfrac{\partial}{\partial x} \end{bmatrix} \overline{\boldsymbol{M}}^{el}$$

$$= \begin{bmatrix} \dfrac{\partial \omega^{el}}{\partial x} M_1 + \omega^{el} \dfrac{\partial M_1}{\partial x} & 0 & \cdots & \dfrac{\partial \omega^{el}}{\partial x} M_{n \times m} + \omega^{el} \dfrac{\partial M_{n \times m}}{\partial x} & 0 \\[4mm] 0 & \dfrac{\partial \omega^{el}}{\partial y} M_1 + \omega^{el} \dfrac{\partial M_1}{\partial y} & \cdots & 0 & \dfrac{\partial \omega^{el}}{\partial y} M_{n \times m} + \omega^{el} \dfrac{\partial M_{n \times m}}{\partial y} \\[4mm] \dfrac{\partial \omega^{el}}{\partial y} M_1 + \omega^{el} \dfrac{\partial M_1}{\partial y} & \dfrac{\partial \omega^{el}}{\partial x} M_1 + \omega^{el} \dfrac{\partial M_1}{\partial x} & \cdots & \dfrac{\partial \omega^{el}}{\partial y} M_{n \times m} + \omega^{el} \dfrac{\partial M_{n \times m}}{\partial y} & \dfrac{\partial \omega^{el}}{\partial x} M_{n \times m} + \omega^{el} \dfrac{\partial M_{n \times m}}{\partial x} \end{bmatrix}$$

$$(4.37)$$

Notice that the displacement field can also be weighted to impose DBCs along each direction. In this case, two weighting functions can be introduced into $\overline{\boldsymbol{M}}^{el}$ so that

$$\overline{\boldsymbol{M}}^{el} = \begin{bmatrix} \omega_x M_1 & 0 & \cdots & \omega_x M_{n \times m} & 0 \\[2mm] 0 & \omega_y M_1 & \cdots & 0 & \omega_y M_{n \times m} \end{bmatrix} \qquad (4.38)$$

By definition, HDBCs can be realized by imposing $r^h = 0$ and $\boldsymbol{g}^h = \boldsymbol{0}$. To clarify the idea, a 1D example of the weighting function $\omega(x)$ is illustrated in Fig. 4.5. Here, the embedding domain D is denoted by $[x_A, x_B]$. The physical domain Ω is (x_C, x_D) and the fixed Dirichlet boundary Γ_D is at point x_C. Therefore, the weighting function $\omega(x_C) = 0$ holds for the homogeneous Dirichlet constraint $u|_{x_C} = 0$; it should also be nonzero and smooth in the domain $(\Omega \cup \partial\Omega)$ $\backslash \Gamma_D$ to keep the effects of basis functions. ω_1 and ω_2 shown in Fig. 4.5 represent two typical forms satisfying these conditions.

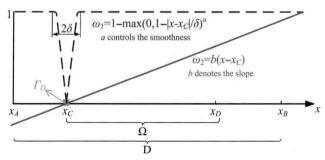

FIG. 4.5 Weighting functions for 1D weighted B-spline.

From Eqs. (4.34), (4.35), it can clearly be seen that the key issue is how to construct the weighting function and boundary value function. To make the discussion easy, both ω^{th} and ω^{el} are denoted by ω below while r^h, g_x^h, and g_y^h are all denoted by G.

4.2.3   Formulations of the weighting function and the boundary value function

Without loss of generality, suppose multiple Dirichlet boundaries numbered Γ_{D1}, Γ_{D2}, ..., Γ_{Dl} are involved in a physical problem with prescribed values

$$u = g_i \quad \text{on } \Gamma_{D_i} \tag{4.39}$$

and each Dirichlet boundary Γ_{Di} is geometrically represented by $\omega_i(x)$ so that

$$\omega_i(x) = 0 \quad \text{only on } \Gamma_{D_i} \tag{4.40}$$

Thus, constructions of the weighting function $\omega(x)$ and the boundary value function G have to satisfy

$$\omega(x) = 0 \quad \forall \omega_i(x) = 0 \tag{4.41}$$

and

$$G(x) = g_i \quad \forall \omega_i(x) = 0 \tag{4.42}$$

In fact, as the weighting function and boundary value function are in the form of level-set functions, the expressions are not unique. Typical forms are given below.

(i) Constructions of weighting function

The level-set functions given in Section 3.3 are used for the definitions of weighting functions. First of all, typical Dirichlet boundaries are presented for a point and a curved segment.

- *Weighting function for the representation of DBC at a point*

If the DBC concerns a point Q (x_0, y_0), the weighting function $\omega(x, y)$ can be constructed as the square of the distance function

$$\omega(x, y) = (x - x_0)^2 + (y - y_0)^2 \tag{4.43}$$

- *Weighting function for the representation of DBC over a segmented curve*

As illustrated in Fig. 4.6, f is the level-set function of an open curve and g is the level-set function of a closed curve. Suppose that DBC is imposed on the segmented curve of f encircled by g. Based on the property of the R-function, a positive weighting function ω related to the segment can be formulated as

$$\omega = -\left(g \wedge \left(-f^2\right)\right) \tag{4.44}$$

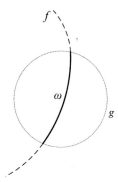

FIG. 4.6 Weighting function related to curve f segmented by a closed curve g.

Example 4.1 Weighting function of the Dirichlet boundary defined by an ellipse cut by a circle

As an example, consider the Dirichlet boundary defined by an ellipse f cut by a circle g, as shown in Fig. 4.7A. Their equations are expressed as

$$
\begin{cases}
f = 1 - \dfrac{x^2}{36} - \dfrac{y^2}{25} \\[2mm]
g = 1 - \dfrac{(x-6)^2}{9} - \dfrac{y^2}{9}
\end{cases}
\tag{4.45}
$$

The weighting function related to the Dirichlet boundary can thus be constructed with the R_0 function as

$$
\omega = -\left(g + h - \sqrt{g^2 + h^2} \right)
\tag{4.46}
$$

with

$$
h = -f^2 = -\left(1 - \frac{x^2}{36} - \frac{y^2}{25} \right)^2
\tag{4.47}
$$

The contour of the weighting function $\omega(x, y)$ is shown in Fig. 4.7B.

- *Weighting function defined with the smooth distance function* (Höllig et al., 2005)

This function is similar to the step function with the merit of reducing the computational complexity because only K and F in the transition strip need revising. It is featured by a plateau of height 1 beyond a fixed distance from the boundary and holds the following form

$$
\omega(\mathbf{x}) = 1 - \max\left(1, \frac{\delta - \text{dist}(\mathbf{x}, \partial\Omega)}{\delta} \right)^{\gamma}
\tag{4.48}
$$

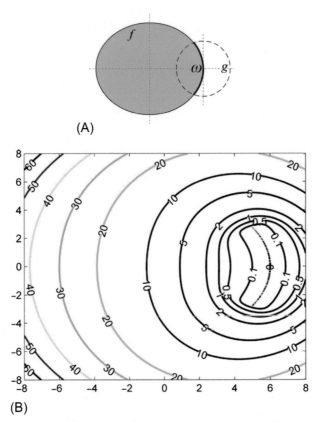

FIG. 4.7 Construction of the weighting function of an open curve: (A) ellipse, circle, and their intersected segment; and (B) iso-contours of the weighting function $\omega(x,y)$.

where dist$(x,\partial\Omega)$ denotes the distance from an arbitrary point x to the physical boundary $\partial\Omega$. δ is a fixed distance parameter, that is, the width of the increasing transition strip. γ is the smoothing parameter controlling the strip continuity.

Fig. 4.8 illustrates this function for the definition of DBC at two points $x=\pm2$ in the 1D case. It is observed that $\omega=0$ holds only at $x=\pm2$ and that $\omega=1$ holds constantly when the point is away from $x=\pm2$ at a minimum distance of δ. The transition strip is enlarged with the increase of δ and smoothed with the increase of γ.

- *Weighting function for multiple Dirichlet boundaries*

The above discussions clearly illustrate how the weighing function is formulated by the level-set function for a single Dirichlet boundary. Concerning multiple Dirichlet boundaries, the weighting function $\omega(x)$ in Eq. (4.41) acts as a multiplier to modify the original interpolation functions and has to vanish on any Dirichlet boundary $(\omega_i(x)=0)$. Here, two construction techniques are

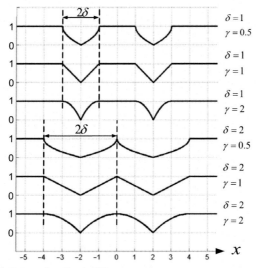

FIG. 4.8 Smooth distance function in 1D case.

presented. First, a straightforward method is to build the weighting function by means of the product

$$\omega(x) = \prod_{i=1}^{l} \omega_i(x) \tag{4.49}$$

Obviously, $\omega(x) = 0$ if any $\omega_i(x) = 0$ on Γ_{Di}. Nevertheless, a product of all the implicit functions might lead to a surge of function values, which will cause a numerical overflow and then result in poor computing accuracy and poor robustness. Therefore, each level-set function $\omega_i(x)$ is preferred to be normalized in advance according to the size of the physical domain. For instance, the level-set function of a circle can be normalized in terms of its radius.

$$\omega(x) = R - \sqrt{(x - x_0)^2 + (y - y_0)^2} \tag{4.50}$$

Besides, according to Höllig et al. (2005), the weighting function can also be constructed with the R-function, the KS function, the Ricci function, or the step function so that

$$\omega(x) = \omega_1 \wedge \omega_2 \wedge \cdots \wedge \omega_l \tag{4.51}$$

When more than two Dirichlet boundaries are involved, the n-ary R-function in Eq. (3.56) can be used once to define the resulting weighting function.

Example 4.2 Weighting function of multiple Dirichlet boundaries
For example, consider a 2D domain bounded by an ellipse, a circle, and a super-ellipse, as depicted in Fig. 4.9. The level-set functions of the three boundaries read

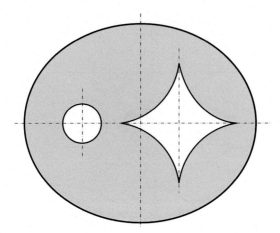

FIG. 4.9 Geometric domain defined by an ellipse, a superellipse, and a circle.

$$\begin{cases} \omega_1(x,y) = 1 - \dfrac{x^2}{36} - \dfrac{y^2}{25} \\ \omega_2(x,y) = (x+3)^2 + y^2 - 1 \\ \omega_3(x,y) = \left|\dfrac{x-2}{3}\right|^{1/2} + \left|\dfrac{y}{3}\right|^{1/2} - 1 \end{cases} \qquad (4.52)$$

Fig. 4.10 shows that both the product function and the R-function can ensure a zero-value assigned to the specified boundaries. However, it is obvious that the overall function value of the former is much greater than the latter.

(ii) Constructions of boundary value function

According to Eqs. (4.39), (4.42), the boundary value function $G(x)$ has to represent imposed IDBCs. Here, a linear combination of g_i is formulated to define $G(x)$

$$G(x) = \sum_{i=1}^{l} b_i(x) g_i \qquad (4.53)$$

where $b_i(x)$ denotes the ith weighting coefficient associated with g_i at Γ_{Di}. According to Eq. (4.53), the basic properties of $b_i(x)$ are as follows

$$b_i(x)|_{\Gamma_{D_j}} = \delta_{ij} \quad i,j = 1,2,\dots,l \qquad (4.54)$$

with δ_{ij} denoting the Kronecker delta function. With this condition, different forms are proposed below for the definition of $b_i(x)$.

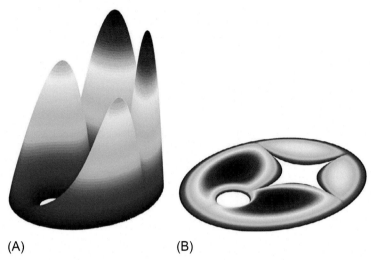

(A)   (B)

FIG. 4.10 Weighting functions of the geometric domain defined by an ellipse, a superellipse, and a circle: (A) weighting function constructed by the product function of Eq. (4.49); and (B) weighting function constructed by Eq. (4.51) with R-function.

- *Transfinite interpolation form*

The weighting coefficients can be defined by extending the transfinite interpolation that was studied by Rvachev et al. (2001) in the CAD community

$$b_i(\mathbf{x}) = \frac{\prod\limits_{j=1, j \neq i}^{l} \omega_j(\mathbf{x})}{\sum\limits_{k=1}^{l} \prod\limits_{j=1, j \neq i}^{l} \omega_j(\mathbf{x})} \qquad (4.55)$$

The above equation shares a property of symmetry and similarity. $\omega_j(\mathbf{x})$ is generally supposed to be positive in Ω to ensure the nonzero value of the denominator. Evidently, $b_i = 1$ only if $\omega_i = 0$ and the partition of unity holds with $\sum_{i=1}^{l} b_i = 1$.

- *General form*

In view of Eq. (4.55), an alternative form is also proposed as

$$b_i(\mathbf{x}) = \frac{\prod\limits_{j=1, j \neq i}^{l} \omega_j(\mathbf{x})}{\omega_i(\mathbf{x}) \cdot c_i(\mathbf{x}) + \prod\limits_{j=1, j \neq i}^{l} \omega_j(\mathbf{x})} \qquad (4.56)$$

where $c_i(x)$ is an arbitrary level-set function defined in Ω. In fact, the weighting coefficients of Eq. (4.56) degenerate into Eq. (4.55) when the following particular form is adopted

$$c_i(x) = \begin{cases} 1 & l=2 \\ \sum\limits_{k=1,k\neq ij=1,j\neq k,i}^{l} \prod\limits_{}^{l} \omega_j(x) & l \geq 3 \end{cases} \tag{4.57}$$

To have a clear idea, consider $l=2$ as an example. Eq. (4.55) gives rise to

$$G = \frac{\omega_2}{\omega_1 + \omega_2} g_1 + \frac{\omega_1}{\omega_1 + \omega_2} g_2 \tag{4.58}$$

And Eq. (4.56) produces

$$G = \frac{\omega_2}{\omega_1 c_1 + \omega_2} g_1 + \frac{\omega_1}{\omega_1 + \omega_2 c_2} g_2 \tag{4.59}$$

The properties of the boundary value function G can easily be checked out to satisfy the following conditions

$$\begin{cases} G|_{\omega_1=0} = \dfrac{\omega_2}{0 \cdot c_1 + \omega_2} g_1 + \dfrac{0}{0 + \omega_2 c_2} g_2 = \dfrac{\omega_2}{\omega_2} g_1 = g_1 \\ G|_{\omega_2=0} = \dfrac{0}{\omega_1 \cdot c_1 + 0} g_1 + \dfrac{\omega_1}{\omega_1 + 0 \cdot c_2} g_2 = \dfrac{\omega_1}{\omega_1} g_2 = g_2 \end{cases} \tag{4.60}$$

- *General high-order form*

Based on Eq. (4.56), the weighting coefficients and boundary value function can further be generalized as

$$b_i(x) = \frac{\prod\limits_{j=1,j\neq i}^{l} \omega_j^\lambda(x)}{\omega_i^\lambda(x) \cdot c_i(x) + \prod\limits_{j=1,j\neq i}^{l} \omega_j^\lambda(x)} \tag{4.61}$$

Eq. (4.61) indicates that the general high-order form changes the order and the sign of $\omega_i(x)$ to effectively avoid the zero value of the denominator when λ is an even number and $c_i(x) > 0$. This provides a possible way to abrogate the positive condition of $\omega_i(x)$.

As a matter of fact, all the above three forms can be used to construct the boundary value function. Notice that only the transfinite interpolation has the property of normalization with the partition of unity, which is, however, unnecessary for the boundary value function. The equations and properties are summarized in Table 4.2.

TABLE 4.2 Comparisons of different boundary value functions.

	Transfinite interpolation (Rvachev et al., 2001)	General form	General high-order form
Weighting coefficients	$b_i = \dfrac{\prod\limits_{j=1,\,j\neq i}^{l}\omega_j}{\sum\limits_{k=1}^{l}\prod\limits_{j=1,\,j\neq k}^{l}\omega_j}$	$b_i = \dfrac{\prod\limits_{j=1,\,j\neq i}^{l}\omega_j}{\omega_i\cdot c_i+\prod\limits_{j=1,\,j\neq k}^{l}\omega_j}$	$b_i = \dfrac{\prod\limits_{j=1,\,j\neq i}^{l}\omega_j^{\lambda}}{\omega_j^{\lambda}\cdot c_i+\prod\limits_{j=1,\,j\neq k}^{l}\omega_j^{\lambda}}$
Boundary value function		$G(x)=\sum\limits_{i=1}^{l}b_i(x)g_i$	
Conditions	(i) $\omega_i(x)=0$ on Γ_{D_i} (ii) $\omega_i(x)>0$ in Ω	(i) $\omega_i(x)=0$ on Γ_{D_i} (ii) $\omega_i(x)>0$ in Ω (iii) $c_i(x)>0$ in Ω	(i) $\omega_i(x)=0$ on Γ_{D_i} (ii) $c_i(x)>0$ in Ω
Properties	$\sum\limits_{i=1}^{l}b_i=1$	$\sum\limits_{i=1}^{l}b_i\neq1$	$\sum\limits_{i=1}^{l}b_i\neq1$

Example 4.3 Boundary value function of multiple Dirichlet boundaries
To clarify the idea, consider the geometric domain shown in Fig. 4.9. Suppose the prescribed values of g_1, g_2, and g_3 are imposed to the boundary of the ellipse, the inner zone of the circle, and the inner zone of the superellipse defined by the zero level-sets of implicit functions ω_1, ω_2, and ω_3, respectively. The involved functions are expressed as

$$\begin{cases} \omega_1(x,y) = 1-\dfrac{x^2}{36}-\dfrac{y^2}{25} \\[2mm] \omega_2(x,y) = (x+3)^2+y^2-1+\left|(x+3)^2+y^2-1\right| \\[2mm] \omega_3(x,y) = \left|\dfrac{x-2}{3}\right|^{1/2}+\left|\dfrac{y}{3}\right|^{1/2}-1+\left|\left|\dfrac{x-2}{3}\right|^{1/2}+\left|\dfrac{y}{3}\right|^{1/2}-1\right| \\[2mm] g_1=0,\ \ g_2=1,\ \ g_3=3 \end{cases} \tag{4.62}$$

Boundary value functions are constructed in four different forms. Eq. (4.55) produces

$$G_1 = \frac{\omega_2\omega_3}{\omega_1\omega_2+\omega_2\omega_3+\omega_1\omega_3}g_1+\frac{\omega_1\omega_3}{\omega_1\omega_2+\omega_2\omega_3+\omega_1\omega_3}g_2+\frac{\omega_1\omega_2}{\omega_1\omega_2+\omega_2\omega_3+\omega_1\omega_3}g_3 \tag{4.63}$$

and the high-order form

$$G_2 = \frac{\omega_2^2 \omega_3^2}{\omega_1^2 \omega_2^2 + \omega_2^2 \omega_3^2 + \omega_1^2 \omega_3^2} g_1 + \frac{\omega_1^2 \omega_3^2}{\omega_1^2 \omega_2^2 + \omega_2^2 \omega_3^2 + \omega_1^2 \omega_3^2} g_2 + \frac{\omega_1^2 \omega_2^2}{\omega_1^2 \omega_2^2 + \omega_2^2 \omega_3^2 + \omega_1^2 \omega_3^2} g_3$$

(4.64)

Eq. (4.56) produces

$$G_3 = \frac{\omega_2 \omega_3}{\omega_1 + \omega_2 \omega_3} g_1 + \frac{\omega_1 \omega_3}{\omega_2 + \omega_1 \omega_3} g_2 + \frac{\omega_1 \omega_2}{\omega_3 + \omega_1 \omega_2} g_3 \quad (c_i = 1)$$

(4.65)

and the high-order form

$$G_4 = \frac{\omega_2^2 \omega_3^2}{\omega_1^2 + \omega_2^2 \omega_3^2} g_1 + \frac{\omega_1^2 \omega_3^2}{\omega_2^2 + \omega_1^2 \omega_3^2} g_2 + \frac{\omega_1^2 \omega_2^2}{\omega_3^2 + \omega_1^2 \omega_2^2} g_3 \quad (c_i = 1)$$

(4.66)

where G_1 and G_3 correspond to the transfinite interpolation and general form, respectively. G_2 and G_4 refer to the square forms of them, respectively. Fig. 4.11 presents four different forms of the boundary value function in the physical domain.

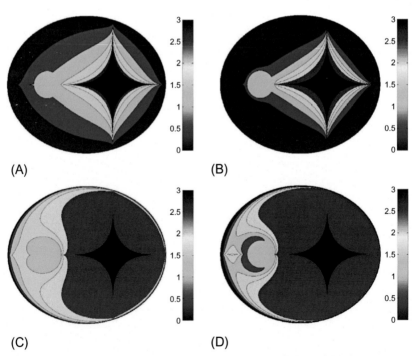

(A) (B)

(C) (D)

FIG. 4.11 Different boundary value functions: (A) G_1; (B) G_2; (C) G_3; and (D) G_4.

4.3 Numerical examples

In this section, four problems are investigated to illustrate the effectiveness of the weighted B-spline FCM. The relative errors of temperature T, displacement $\boldsymbol{u}$, and stress $\boldsymbol{\sigma}$ in the L^2-norm are adopted to measure the numerical precision.

$$
\begin{cases}
e_T = \dfrac{\|T - T^h\|_{L^2,\Omega}}{\|T\|_{L^2,\Omega}} = \sqrt{\dfrac{\displaystyle\int_D (T - T^h)^2 \alpha d\Omega}{\displaystyle\int_D T^2 \alpha d\Omega}} \\[4ex]
e_{\boldsymbol{u}} = \dfrac{\|\boldsymbol{u} - \boldsymbol{u}^h\|_{L^2,\Omega}}{\|\boldsymbol{u}\|_{L^2,\Omega}} = \sqrt{\dfrac{\displaystyle\int_D (\boldsymbol{u} - \boldsymbol{u}^h)^T \alpha (\boldsymbol{u} - \boldsymbol{u}^h) d\Omega}{\displaystyle\int_D \boldsymbol{u}^T \alpha \boldsymbol{u} d\Omega}} \\[4ex]
e_{\boldsymbol{\sigma}} = \dfrac{\|\boldsymbol{\sigma} - \boldsymbol{\sigma}^h\|_{L^2,\Omega}}{\|\boldsymbol{\sigma}\|_{L^2,\Omega}} = \sqrt{\dfrac{\displaystyle\int_D (\boldsymbol{\sigma} - \boldsymbol{\sigma}^h)^T \alpha (\boldsymbol{\sigma} - \boldsymbol{\sigma}^h) d\Omega}{\displaystyle\int_D \boldsymbol{\sigma}^T \alpha \boldsymbol{\sigma} d\Omega}}
\end{cases} \tag{4.67}
$$

in which T^h, $\boldsymbol{u}^h$, and $\boldsymbol{\sigma}^h$ are the corresponding numerical solutions related to the mesh discretization size h.

4.3.1 Infinite plate with a circular hole

This problem is shown in Fig. 4.12A. It is an infinite plate with a circular hole and loaded by a uniform traction $q = 10\,\text{MPa}$ on both ends. The Young's modulus and Poisson's ratio are $200\,\text{GPa}$ and 0.29, respectively.

Theoretically, analytic stress solutions (Schillinger et al., 2012a; Boresi et al., 1985) exist in the polar coordinate system

$$
\begin{cases}
\sigma_r(r, \theta) = \dfrac{q}{2}\left(1 - \dfrac{R^2}{r^2}\right) + \dfrac{q}{2}\left(1 - \dfrac{R^2}{r^2}\right)\left(1 - \dfrac{3R^2}{r^2}\right)\cos 2\theta \\[3ex]
\sigma_\theta(r, \theta) = \dfrac{q}{2}\left(1 + \dfrac{R^2}{r^2}\right) - \dfrac{q}{2}\left(1 + \dfrac{3R^4}{r^4}\right)\cos 2\theta \\[3ex]
\tau_{r\theta}(r, \theta) = -\dfrac{q}{2}\left(1 - \dfrac{R^2}{r^2}\right)\left(1 + \dfrac{3R^2}{r^2}\right)\sin 2\theta
\end{cases} \tag{4.68}
$$

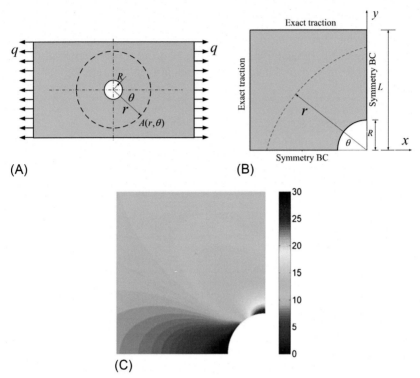

FIG. 4.12 An infinite plate with a circular hole ($R = 1$ mm, $L = 4$ mm): (A) original model; (B) simplified model; and (C) analytic solution of σ_x.

The stress tensor can further be transformed into the Cartesian coordinate system

$$\left\{\begin{array}{c} \sigma_x \\ \sigma_y \\ \tau_{xy} \end{array}\right\} = \left[\begin{array}{ccc} \cos^2\theta & \sin^2\theta & -\sin 2\theta \\ \sin^2\theta & \cos^2\theta & \sin 2\theta \\ \sin\theta\cos\theta & -\sin\theta\cos\theta & \cos 2\theta \end{array}\right] \left\{\begin{array}{c} \sigma_r \\ \sigma_\theta \\ \tau_{r\theta} \end{array}\right\} \quad (4.69)$$

Fig. 4.12C indicates that the maximum stress value occurs at the top point ($R = 1$ mm, $\theta = 90°$) of the hole.

The weighted B-spline FCM is now used. Only a quarter of the structure shown in Fig. 4.12B is considered as a simplified model due to the symmetry and embedded into a square domain ($L \times L$). The latter is further discretized into 32×32 finite cells. Besides, each boundary cell is partitioned five times using the quadtree refinement, as illustrated in Fig. 4.13. Cubic B-spline basis functions are adopted as interpolation functions.

Considering the symmetry, two linear weighting functions ω_x and ω_y are used to apply DBC along the x and y directions at $x = 0$ and $y = 0$, respectively.

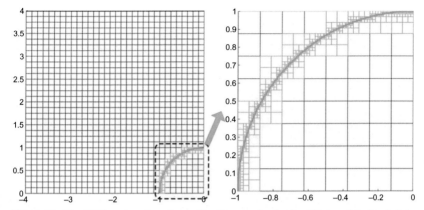

FIG. 4.13 The FCM mesh and quadtree cell refinement for the simplified model of an infinite plate with a circular hole.

$$\begin{cases} \omega_x = x \\ \omega_y = y \end{cases} \tag{4.70}$$

The computing result of σ_x is shown in Fig. 4.14A. The smooth distance functions given below are further tested

$$\begin{cases} \omega_x = 1 - \max\left(0, 1 - \dfrac{|x|}{\delta}\right)^{\gamma} \\ \omega_y = 1 - \max\left(0, 1 - \dfrac{|y|}{\delta}\right)^{\gamma} \end{cases} \tag{4.71}$$

The stress results are shown in Fig. 4.14B and C for different values of the parameters δ and γ.

The pointwise relative error of σ_x along $x=0$ is illustrated in Fig. 4.15 in order to check the accuracy of the weighted B-spline FCM. As the cubic B-spline has a good continuty, the weighting functions almost do not influence the stress result. The maximal relative error of σ_x along the y axis is always smaller than 0.6%.

4.3.2 A cylindrical sector subjected to harmonic Dirichlet boundary conditions

This example was originally studied using the isogeometric analysis in Chen et al. (2012). Fig. 4.16A shows that the inner and outer radius are $R_1=1$ and $R_2=2$, respectively. To make a dimensionless study, Young's modulus E and Poisson's ratio μ are set to 1 and 0.3, respectively. Suppose a displacement

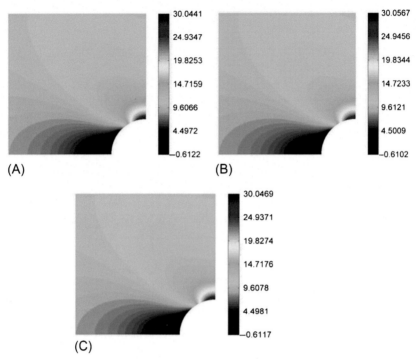

FIG. 4.14 σ_x obtained with linear weighting function and smooth distance function: (A) linear weighting function; (B) smooth distance function with $\delta = 1$, $\gamma = 3$; and (C) smooth distance function with $\delta = 4$, $\gamma = 3$.

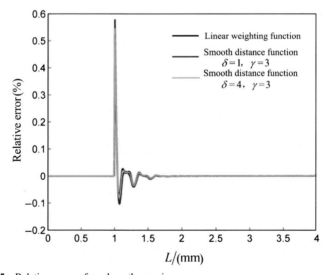

FIG. 4.15 Relative errors of σ_x along the y axis.

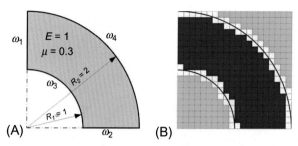

FIG. 4.16 The schematic model and FCM mesh of the cylindrical sector: (A) schematic model; and (B) FCM mesh.

constraint $u_x = u_y = \sin(\pi x)\sin(\pi y)$ is imposed as an IDBC along the whole boundary and a body force is expressed as

$$f_x = f_y = \frac{E\pi^2}{1-\mu^2} \left(\frac{3-\mu}{2} \sin(\pi x)\sin(\pi y) - \frac{1+\mu}{2} \cos(\pi x)\cos(\pi y) \right) \quad (4.72)$$

According to Chen et al. (2012), the analytical solutions of displacement and stress are formulated as

$$u_x = u_y = \sin(\pi x)\sin(\pi y) \quad (4.73)$$

$$\sigma = \frac{E\pi}{1-\mu^2} \cdot \begin{bmatrix} \cos(\pi x)\sin(\pi y) + \mu \sin(\pi x)\cos(\pi y) \\ \mu \cos(\pi x)\sin(\pi y) + \sin(\pi x)\cos(\pi y) \\ (1-\mu)\sin\pi(x+y)/2 \end{bmatrix} \quad (4.74)$$

Fig. 4.17 gives the analytical solutions of x-direction displacement, x-direction normal stress σ_x, and von Mises stress. Based on the FCM, the model is embedded into a 2×2 square domain that is further discretized into 20×20 cells, as shown in Fig. 4.16B. Each boundary cell is treated by five-level refinements using the quadtree approach in Fig. 4.3 so as to conduct the Gaussian integration with good accuracy. The cubic B-spline basis functions are adopted as interpolation functions. The IDBC is imposed using the proposed method in Section 4.2.2. The level-set functions of the four boundaries in Fig. 4.16A can easily be defined as

$$\begin{cases} \omega_1 = x \\ \omega_2 = y \\ \omega_3 = \sqrt{x^2 + y^2} - 1 \\ \omega_4 = 2 - \sqrt{x^2 + y^2} \end{cases} \quad (4.75)$$

According to Eq. (4.49), the weighting function of the analytic product can be obtained in a straightforward way

$$\omega = \omega_1 \cdot \omega_2 \cdot \omega_3 \cdot \omega_4 \quad (4.76)$$

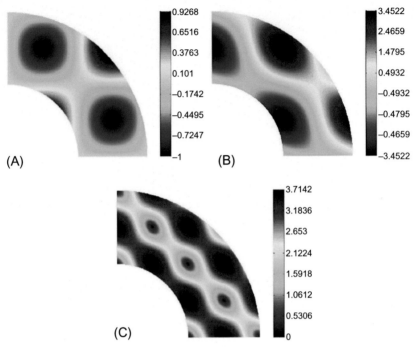

FIG. 4.17 Analytical solutions over the cylindrical sector: (A) x-direction displacement; (B) normal stress σ_x; and (C) von Mises stress.

According to Table 4.2, the transfinite interpolation, the general form, and the general high-order form are applied to define the boundary value functions that are given in Eq. (4.77). Fig. 4.18 depicts these boundary value functions and the corresponding boundary values along the whole boundary of the model, which indicates that they have different distributions but share the same boundary value along the whole boundary. Especially, G_1 related to the transfinite interpolation corresponds to the analytical solution given in Eq. (4.73).

$$
\begin{cases}
G_1 = \dfrac{\omega_2\omega_3\omega_4 \cdot g_1 + \omega_1\omega_3\omega_4 \cdot g_2 + \omega_1\omega_2\omega_4 \cdot g_3 + \omega_1\omega_2\omega_3 \cdot g_4}{\omega_2\omega_3\omega_4 + \omega_1\omega_3\omega_4 + \omega_1\omega_2\omega_4 + \omega_1\omega_2\omega_3} \sin(\pi x)\sin(\pi y) \\[2ex]
G_2 = \left(\dfrac{\omega_2\omega_3\omega_4}{\omega_1 + \omega_2\omega_3\omega_4} + \dfrac{\omega_1\omega_3\omega_4}{\omega_2 + \omega_1\omega_3\omega_4} + \dfrac{\omega_1\omega_2\omega_4}{\omega_3 + \omega_1\omega_2\omega_4} + \dfrac{\omega_1\omega_2\omega_3}{\omega_4 + \omega_1\omega_2\omega_3} \right) \sin(\pi x)\sin(\pi y) \\[2ex]
G_3 = \left(\dfrac{\omega_2^2\omega_3^2\omega_4^2}{\omega_1^2 + \omega_2^2\omega_3^2\omega_4^2} + \dfrac{\omega_1^2\omega_3^2\omega_4^2}{\omega_2^2 + \omega_1^2\omega_3^2\omega_4^2} + \dfrac{\omega_1^2\omega_2^2\omega_4^2}{\omega_3^2 + \omega_1^2\omega_2^2\omega_4^2} + \dfrac{\omega_1^2\omega_2^2\omega_3^2}{\omega_4^2 + \omega_1^2\omega_2^2\omega_3^2} \right) \sin(\pi x)\sin(\pi y)
\end{cases}
$$

$$(4.77)$$

with

$$g_1 = g_2 = g_3 = g_4 = \sin(\pi x)\sin(\pi y) \qquad (4.78)$$

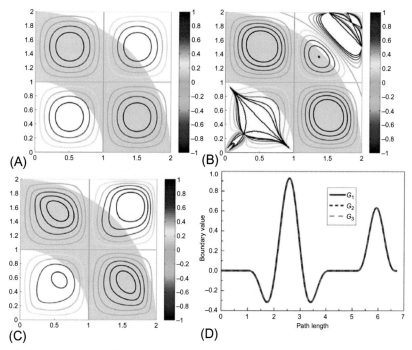

FIG. 4.18 Comparisons of different boundary value functions in Eq. (4.77): (A) G_1; (B) G_2; (C) G_3; and (D) boundary value along the boundary (anticlockwise starting from the point $(1, 0)$).

Figs. 4.19–4.21 depict numerical solutions and corresponding absolute errors using the transfinite interpolation, the general form, and the general high-order form, respectively. All the results are obtained by using 500×500 sampling points extracted from the embedding domain. Obviously, the absolute errors of the x-direction displacement related to the three forms of boundary value functions in Eq. (4.77) are as small as about 10^{-3}. It is worth mentioning that the absolute errors of displacement along the whole boundary are actually zero owing to the exact imposition of boundary value functions.

Fig. 4.22A–C shows the relations between e_u and h-refinement using different boundary value functions while Fig. 4.22D–F refers to the relations between the e_σ and h-refinement, correspondingly. However, Fig. 4.22A and D is quite different for their convergence behavior. The high-order B-spline basis functions result in poor approximations in the case of transfinite interpolation. In fact, the boundary value function G_1 is exactly the analytical solution of the displacement field in Eq. (4.73), so the vector U in Eq. (4.23) should be a zero vector in order to get the exact solution. As is seen from the basic formulas in Eq. (4.23), the interpolated displacement u is no longer polynomial but rational and nonlinear due to the involvement of the weighting functions and boundary value functions in Section 4.2.3. Sometimes, this will cause difficulty in passing

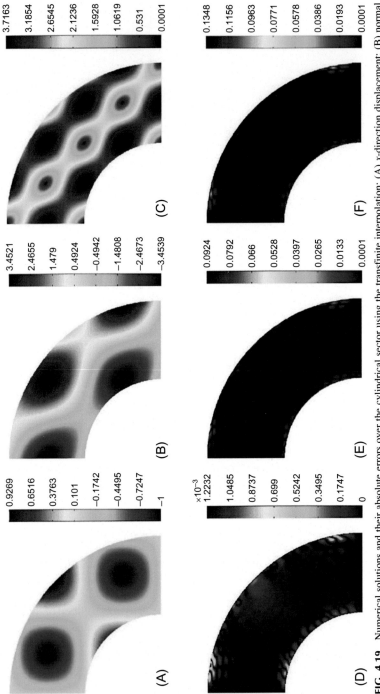

FIG. 4.19 Numerical solutions and their absolute errors over the cylindrical sector using the transfinite interpolation: (A) x-direction displacement; (B) normal stress σ_x; (C) von Mises stress; (D) absolute error of x-direction displacement; (E) absolute error of normal stress σ_x; and (F) absolute error of von Mises stress.

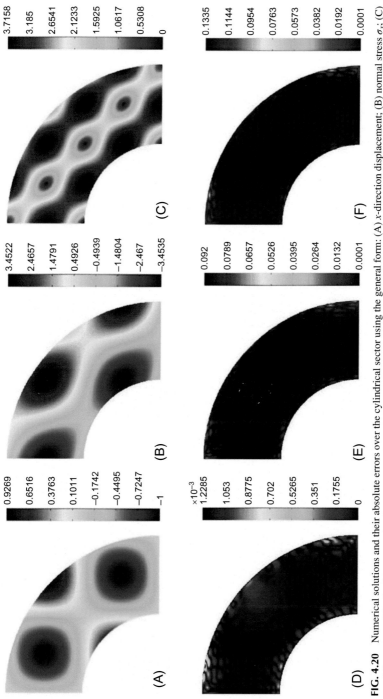

FIG. 4.20 Numerical solutions and their absolute errors over the cylindrical sector using the general form: (A) x-direction displacement; (B) normal stress σ_x; (C) von Mises stress; (D) absolute error of x-direction displacement; (E) absolute error of normal stress σ_x; and (F) absolute error of von Mises stress.

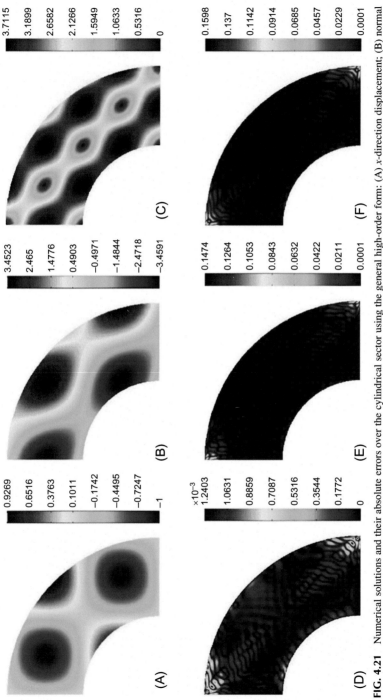

FIG. 4.21 Numerical solutions and their absolute errors over the cylindrical sector using the general high-order form: (A) x-direction displacement; (B) normal stress σ_x; (C) von Mises stress; (D) absolute error of x-direction displacement; (E) absolute error of normal stress σ_x; and (F) absolute error of von Mises stress.

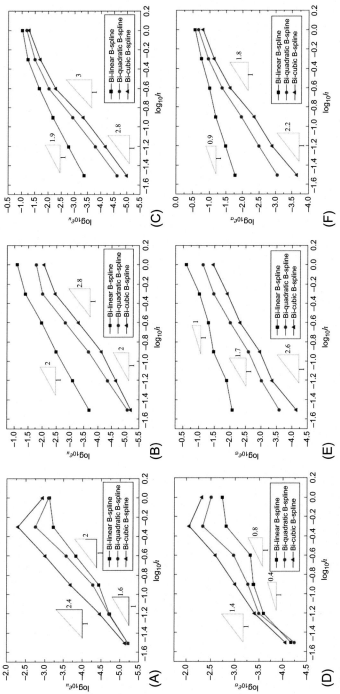

FIG. 4.22 Convergence comparisons with B-spline shape functions of different orders versus h-refinement for the relative error in the L^2-norm using different forms of boundary value function: (A) e_u versus h-refinement using transfinite interpolation; (B) e_u versus h-refinement using general form; (C) e_u versus h-refinement using general high-order form; (D) e_σ versus h-refinement using transfinite interpolation; (E) e_σ versus h-refinement using general form; and (F) e_σ versus h-refinement using general high-order form.

the patch test (Bazeley et al., 1965) and will influence the convergence rate of the proposed method. This might be the reason that the convergence rate related to the cubic B-spline basis function is not always higher than the quadratic one. However, this observation should be further highlighted and investigated theoretically, which can be considered an open question to reveal the underlying mechanism. Nevertheless, Fig. 4.22 also indicates that the error of the numerical approximation is decreasing with the increasing grid density, which means the numerical solutions converge toward the analytical solution.

4.3.3 A cylinder subjected to prescribed radial displacement and temperature

Fig. 4.23 shows that radial displacement and temperature are prescribed along the inner and outer boundaries with $u_{r1}=0.25$, $T_1=3°$ and $u_{r2}=0$, $T_2=1°$, respectively. All the material properties needed are directly given to omit the dimensions. Analytical solutions (Sadd, 2009; Zander et al., 2012) in the polar coordinate system are expressed as

$$T(r) = 1 - \frac{\ln r}{\ln 2} \tag{4.79}$$

$$\begin{cases} u_r(r) = -\dfrac{r}{2}\dfrac{\ln r}{\ln 2} \\ u_\theta = 0 \end{cases} \tag{4.80}$$

$$\begin{cases} \sigma_r(r) = \dfrac{\ln r - 1}{2\ln 2} - 1 \\ \sigma_\theta(r) = \dfrac{\ln r}{2\ln 2} - 1 \\ \tau_{r\theta} = 0 \end{cases} \tag{4.81}$$

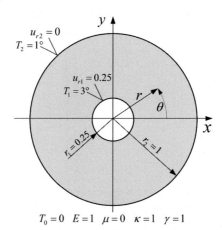

$$T_0 = 0 \quad E = 1 \quad \mu = 0 \quad \kappa = 1 \quad \gamma = 1$$

FIG. 4.23 The schematic model of the cylinder (Zander et al., 2012).

All the above solutions can be transformed into the Cartesian coordinate system with the help of the following relationships

$$\begin{Bmatrix} u \\ v \end{Bmatrix} = \begin{bmatrix} \cos\theta & -\sin\theta \\ \sin\theta & \cos\theta \end{bmatrix} \begin{Bmatrix} u_r \\ u_\theta \end{Bmatrix} \tag{4.82}$$

$$\begin{Bmatrix} \sigma_x \\ \sigma_y \\ \tau_{xy} \end{Bmatrix} = \begin{bmatrix} \cos^2\theta & \sin^2\theta & -\sin 2\theta \\ \sin^2\theta & \cos^2\theta & \sin 2\theta \\ \sin\theta\cos\theta & -\sin\theta\cos\theta & \cos 2\theta \end{bmatrix} \begin{Bmatrix} \sigma_r \\ \sigma_\theta \\ \tau_{r\theta} \end{Bmatrix} \tag{4.83}$$

The finite cell model is constructed by embedding the cylinder into a 2×2 square domain and then discretizing the latter into 8×8 cells. As depicted in Fig. 4.24, the boundary cells marked in white are partitioned into subcells for the sake of numerical integration. Here, the level-set functions of the Dirichlet boundaries are expressed as

$$\begin{cases} \omega_1 = \sqrt{x^2 + y^2} - 0.25 \\ \omega_2 = 1 - \sqrt{x^2 + y^2} \end{cases} \tag{4.84}$$

The weighting function can thus be defined as

$$\omega^{\text{th}} = \omega^{\text{el}} = \omega_1 \cdot \omega_2 \tag{4.85}$$

When transfinite interpolation is used, the boundary value functions applied for both the thermal and elastic fields are written as

$$G_r = \frac{\omega_2}{\omega_1 + \omega_2} \cdot 3 + \frac{\omega_1}{\omega_1 + \omega_2} \cdot 1 \tag{4.86}$$

$$G_{gx} = \frac{\omega_2}{\omega_1 + \omega_2} \cdot \frac{0.25x}{\sqrt{x^2 + y^2}} \tag{4.87}$$

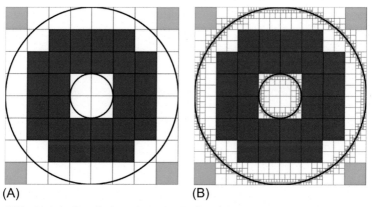

(A)   (B)

FIG. 4.24 Domain discretization: (A) 8×8 cells discretization; and (B) boundary cells with five-level refinements.

$$G_{gy} = \frac{\omega_2}{\omega_1 + \omega_2} \cdot \frac{0.25y}{\sqrt{x^2 + y^2}} \tag{4.88}$$

Fig. 4.25 shows the contour plots of the weighting function and the boundary value function for the temperature field. The former vanishes on the boundary while the latter exactly satisfies the DBCs. The boundary value functions for the displacement field are also plotted in Fig. 4.26. To compare, analytical solutions and numerical solutions using the weighted B-spline FCM are depicted in Figs. 4.27 and 4.28, respectively. We can observe that the distributions of

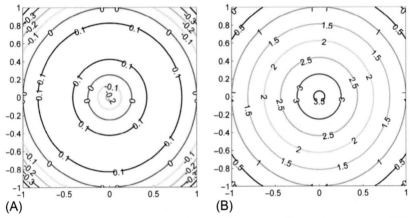

FIG. 4.25 Contour plots of weighting function and boundary value function for temperature field: (A) weighting function; and (B) boundary value function.

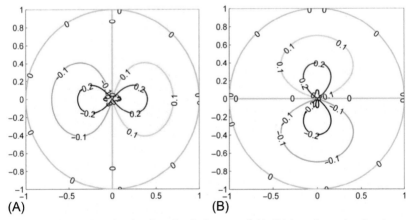

FIG. 4.26 Boundary value functions for displacement field: (A) boundary value function on x-direction; and (B) boundary value function on y-direction.

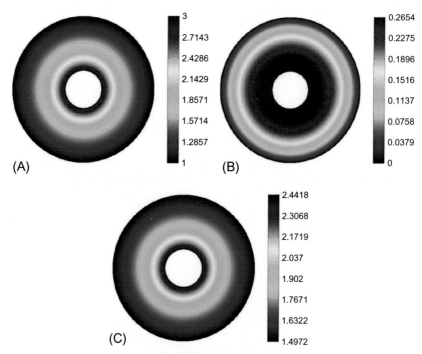

FIG. 4.27 Analytical solutions: (A) analytical temperature distribution; (B) analytical displacement distribution; and (C) analytical von Mises stress distribution.

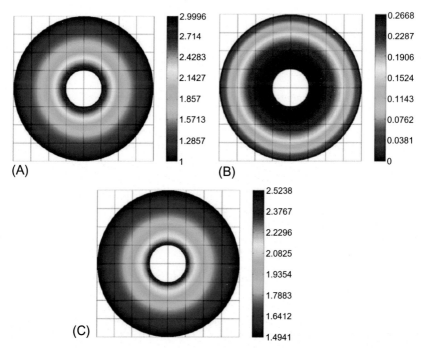

FIG. 4.28 Numerical solutions: (A) numerical temperature distribution; (B) numerical displacement distribution; and (C) numerical von Mises stress distribution.

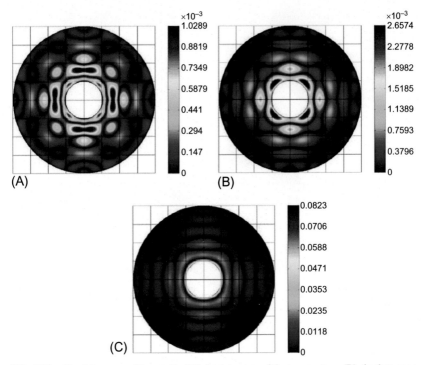

FIG. 4.29 Absolute errors of the results: (A) absolute error of the temperature; (B) absolute error of the displacement; and (C) absolute error of the von Mises stress.

temperature, displacement, and von Mises stress match very well with the analytical solutions. Fig. 4.29 shows that the maximum absolute error of temperature is 1.0289×10^{-3} while those of displacement and von Mises stress are 2.6574×10^{-3} and 0.0823, respectively. Fig. 4.30 shows the convergence comparisons with the B-spline shape functions of different orders against h-refinement in terms of the error measures using Eq. (4.67). It is observed that errors tend to reduce with the mesh refinement. To some extent, the rational form of the boundary value functions influences the convergence behavior, especially for a relatively coarse mesh.

4.3.4 Thermoelastic stress analysis of a heat exchanging device with prescribed temperature

This simplified heat exchanging device, inspired by the work of Zander et al. (2012), was considered here by the proposed method. As depicted in Fig. 4.31, constant temperatures of 35°C and 60°C are imposed on the outer

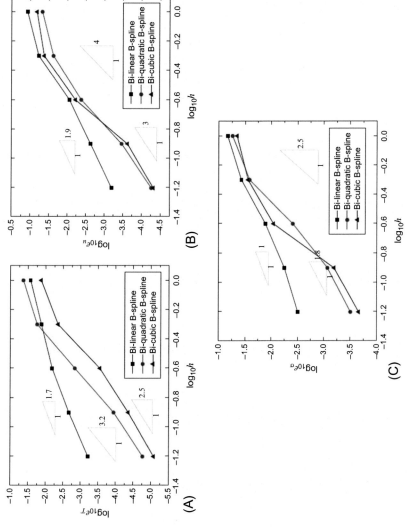

FIG. 4.30 Convergence comparisons with B-spline basis functions of different orders: (A) e_T versus h-refinement; (B) e_u versus h-refinement; and (C) e_σ versus h-refinement.

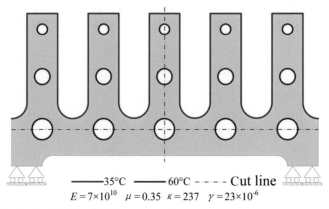

$$E = 7 \times 10^{10} \quad \mu = 0.35 \quad \kappa = 237 \quad \gamma = 23 \times 10^{-6}$$

FIG. 4.31 The simplified model of heat exchanging device.

and inner hole boundaries, respectively. A simple support is also applied on the bottom to fix the vertical displacement.

By means of the weighted B-spline FCM, the model is embedded into a box domain with a discretization of 40×20 finite cells. To ensure the computing accuracy, bi-cubic B-spline basis functions are used and boundary cells are partitioned using five-level quadtree refinements. Based on the Boolean operation given in Fig. 4.32A, the level-set function of the heat exchanging device is constructed using the KS function, as depicted in Fig. 4.32B.

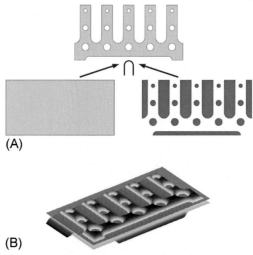

(A)

(B)

FIG. 4.32 The construction and representation of the heat exchange device model: (A) modeling heat exchange device with Boolean operations; and (B) level-set function constructed by the KS function.

As expected, the weighting function and the boundary value function for the temperature field are directly constructed by means of the KS function and Eq. (4.58). Furthermore, the corresponding contour plots are shown in Fig. 4.33, from which it is seen that the weighting function vanishes on each Dirichlet boundary while the boundary value function exactly accords with the assigned boundary values of temperature.

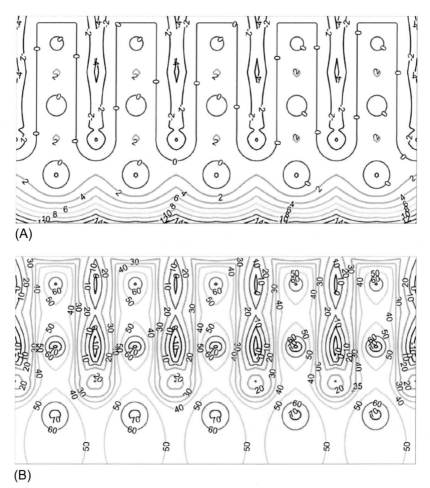

(A)

(B)

FIG. 4.33 Contour plots of the weighting function and boundary value function of the heat exchange device model: (A) weighting function; and (B) boundary value function.

The traditional FEM is also applied for a comparative study. As depicted in Fig. 4.34, the same model is discretized into a conformal mesh of 8456 finite elements that are over 10 times more than finite cells. Fig. 4.35 gives the distributions of temperature and von Mises stress obtained by the weighted B-spline FCM while Fig. 4.36 shows the FEM results. Both kinds of results agree well. The temperature distribution holds the IDBC in Fig. 4.35A thanks to the weighting function and boundary value function. Furthermore, Fig. 4.37 shows the comparison of the different solutions along the horizontal and vertical cut lines in Fig. 4.31 and indicates that the weighted B-spline FCM results match well with the FEM results in the physical domain and even smooth at the interfaces of the solid and void phases.

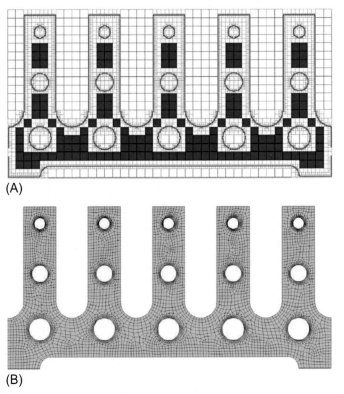

(A)

(B)

FIG. 4.34 Domain discretization with different kinds of meshes: (A) nonconformal mesh of the FCM; and (B) conformal mesh of the FEM.

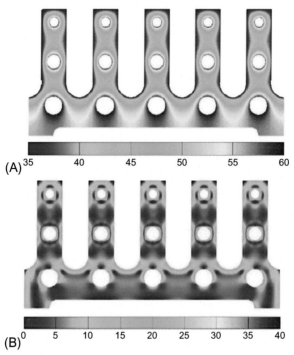

FIG. 4.35 Numerical results using the weighted B-spline FCM: (A) temperature distribution; and (B) von Mises stress distribution.

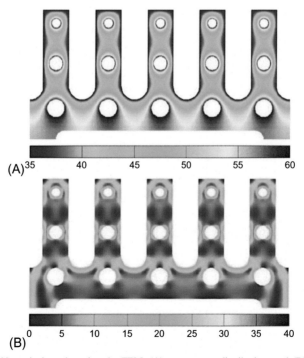

FIG. 4.36 Numerical results using the FEM: (A) temperature distribution; and (B) von Mises stress distribution.

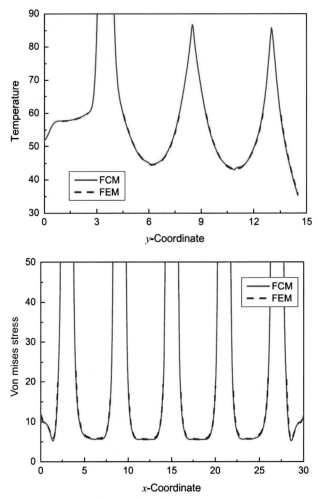

FIG. 4.37   Comparison of the temperature and von Mises stress solutions along the cut lines: (A) temperature along the vertical cut line; and (B) von Mises stress along the horizontal cut line.

References

Abedian, A., Parvizian, J., Düster, A., Rank, E., 2013. The finite cell method for the J2 flow theory of plasticity. Finite Elem. Anal. Des. 69, 37–47.

Ainsworth, M., 2001. Essential boundary conditions and multi-point constraints in finite element analysis. Comput. Methods Appl. Mech. Eng. 190 (48), 6323–6339.

Bazeley, G.P., Cheung, Y.K., Irons, B.M., 1965. Triangular elements in plate bending-conforming and non-conforming solutions (stiffness characteristics of triangular plate elements in bending, and solutions). In: Conference on Matrix Methods in Structural Mechanics, Air Force Institute of Technology. Wright-Patterson, Ohio.

Boresi, A.P., Schmidt, R.J., Sidebottom, O.M., 1985. Advanced Mechanics of Materials. vol. 6. Wiley, New York et al.

Boufflet, J.P., Dambrine, M., Dupire, G., Villon, P., 2012. On the necessity of Nitsche term. Part II: an alternative approach. Appl. Numer. Math. 62 (5), 521–535.

Burla, R.K., Kumar, A.V., 2008. Implicit boundary method for analysis using uniform B-spline basis and structured grid. Int. J. Numer. Methods Eng. 76 (13), 1993–2028.

Chen, T., Mo, R., Wan, N., Gong, Z.W., 2012. Imposing displacement boundary conditions with Nitsche's method in isogeometric analysis. Chin. J. Theor. Appl. Mech. 44 (2), 371–381.

Daux, C., Moës, N., Dolbow, J., Sukumar, N., Belytschko, T., 2000. Arbitrary branched and intersecting cracks with the extended finite element method. Int. J. Numer. Methods Eng. 48 (12), 1741–1760.

Del Pino, S., Pironneau, O., 2003. A fictitious domain based general PDE solver. In: Kuznetsov, Y., Neittanmaki, P., Pironneau, O. (Eds.), Numerical Methods for Scientific Computing: Variational Problems and Applications. CIMNE, Barcelona.

Dupire, G., Boufflet, J.P., Dambrine, M., Villon, P., 2010. On the necessity of Nitsche term. Appl. Numer. Math. 60 (9), 888–902.

Düster, A., Parvizian, J., Yang, Z., Rank, E., 2008. The finite cell method for three-dimensional problems of solid mechanics. Comput. Methods Appl. Mech. Eng. 197 (45–48), 3768–3782.

Embar, A., Dolbow, J., Harari, I., 2010. Imposing Dirichlet boundary conditions with Nitsche's method and spline-based finite elements. Int. J. Numer. Methods Eng. 83 (7), 877–898.

Fernández-Méndez, S., Huerta, A., 2004. Imposing essential boundary conditions in mesh-free methods. Comput. Methods Appl. Mech. Eng. 193 (12–14), 1257–1275.

Glowinski, R., Kuznetsov, Y., 2007. Distributed Lagrange multipliers based on fictitious domain method for second order elliptic problems. Comput. Methods Appl. Mech. Eng. 196 (8), 1498–1506.

Glowinski, R., Pan, T.W., Hesla, T.I., Joseph, D.D., 1999. A distributed Lagrange multiplier/fictitious domain method for particulate flows. Int. J. Multiphase Flow 25 (5), 755–794.

Guo, X., Zhang, W.S., Wang, M.Y., Wei, P., 2011. Stress-related topology optimization via level set approach. Comput. Methods Appl. Mech. Eng. 200 (47–48), 3439–3452.

Höllig, K., Reif, U., 2003. Nonuniform web-splines. Comput. Aided Geom. Des. 20 (5), 277–294.

Höllig, K., Reif, U., Wipper, J., 2001. Weighted extended B-spline approximation of Dirichlet problems. SIAM J. Numer. Anal. 39 (2), 442–462.

Höllig, K., Apprich, C., Streit, A., 2005. Introduction to the Web-method and its applications. Adv. Comput. Math. 23 (1–2), 215–237.

Joulaian, M., Düster, A., 2013. Local enrichment of the finite cell method for problems with material interfaces. Comput. Mech. 52 (4), 741–762.

Kumar, A.V., Padmanabhan, S., Burla, R., 2008. Implicit boundary method for finite element analysis using non-conforming mesh or grid. Int. J. Numer. Methods Eng. 74 (9), 1421–1447.

Moës, N., Béchet, E., Tourbier, M., 2006. Imposing Dirichlet boundary conditions in the extended finite element method. Int. J. Numer. Methods Eng. 67 (12), 1641–1669.

Nitsche, J., 1971, July. Über ein Variationsprinzip zur Lösung von Dirichlet-Problemen bei Verwendung von Teilräumen, die keinen Randbedingungen unterworfen sind. In: Abhandlungen aus dem mathematischen Seminar der Universität Hamburg. vol. 36(1). Springer-Verlag, pp. 9–15.

Parvizian, J., Düster, A., Rank, E., 2007. Finite cell method. Comput. Mech. 41 (1), 121–133.

Ramiere, I., Angot, P., Belliard, M., 2007. A fictitious domain approach with spread interface for elliptic problems with general boundary conditions. Comput. Methods Appl. Mech. Eng. 196 (4–6), 766–781.

Ruess, M., Schillinger, D., Bazilevs, Y., Varduhn, V., Rank, E., 2013. Weakly enforced essential boundary conditions for NURBS-embedded and trimmed NURBS geometries on the basis of the finite cell method. Int. J. Numer. Methods Eng. 95 (10), 811–846.

Rvachev, V.L., Sheiko, T.I., Shapiro, V., Tsukanov, I., 2001. Transfinite interpolation over implicitly defined sets. Comput. Aided Geom. Des. 18 (3), 195–220.

Sadd, M.H., 2009. Elasticity: Theory, Applications, and Numerics. Academic Press.

Schillinger, D., Dede, L., Scott, M.A., Evans, J.A., Borden, M.J., Rank, E., Hughes, T.J., 2012a. An isogeometric design-through-analysis methodology based on adaptive hierarchical refinement of NURBS, immersed boundary methods, and T-spline CAD surfaces. Comput. Methods Appl. Mech. Eng. 249, 116–150.

Schillinger, D., Düster, A., Rank, E., 2012b. The hp-d-adaptive finite cell method for geometrically nonlinear problems of solid mechanics. Int. J. Numer. Methods Eng. 89 (9), 1171–1202.

Wang, M.Y., Li, L., 2013. Shape equilibrium constraint: a strategy for stress-constrained structural topology optimization. Struct. Multidiscip. Optim. 47 (3), 335–352.

Wei, P., Wang, M.Y., Xing, X., 2010. A study on X-FEM in continuum structural optimization using a level set model. Comput. Aided Des. 42 (8), 708–719.

Zander, N., Kollmannsberger, S., Ruess, M., Yosibash, Z., Rank, E., 2012. The finite cell method for linear thermoelasticity. Comput. Math. Appl. 64 (11), 3527–3541.

Zhang, W.S., Guo, X., Wang, M.Y., Wei, P., 2013. Optimal topology design of continuum structures with stress concentration alleviation via level set method. Int. J. Numer. Methods Eng. 93 (9), 942–959.

Zhang, W., Zhao, L., Cai, S., 2015. Shape optimization of Dirichlet boundaries based on weighted B-spline finite cell method and level-set function. Comput. Methods Appl. Mech. Eng. 294, 359–383.

Zienkiewicz, O.C., Taylor, R.L., Nithiarasu, P., Zhu, J.Z., 1977. The Finite Element Method. vol. 3. McGraw-Hill, London.

Chapter 5

Feature-based modeling and sensitivity analysis for structural optimization

5.1 Feature-based modeling with level-set functions

Feature-based modeling is well recognized in the CAD community for its enabling functionality using shape features as basic primitives. In feature-based modeling, an engineering structure or a mechanical part can be interpreted as a collection of various shape features such as slots, stiffeners, reinforcing ribs, threads, holes, etc. These features are designable and needed practically in view of their functionality, aesthetics, assemblage, machining facility, or other considerations.

In this chapter, the concepts of freeform design domain modeler (FDDM) and topology variation modeler (TVM) are introduced to construct the feature-based model for structural optimization. The FDDM describes the predefined freeform design domain while the TVM controls its topology change. In this sense, the topology optimization process is interpreted as a Boolean intersection process between the TVM and the FDDM. In detail, the TVM is constructed by level-set functions such as interpolations of basis functions or engineering features whose parameters are designated as design variables. The FDDM can be constructed using the level-set functions of constituting primitives of the freeform design domain. Finally, both modelers undergo Boolean operations to confine optimization in the predefined freeform design domain.

5.1.1 Freeform design domain modeler

Fig. 5.1 shows a predefined freeform design domain with prescribed intrinsic features. Suppose the intrinsic features are attributed with a specific layer thickness t and remain unchanged during the optimization process on account of their specific functionality. The rest of the design domain is topologically optimized.

For the freeform design domain depicted in Fig. 5.1, the FDDM consists of two submodelers, i.e., a basic modeler of freeform design domain and an

The Feature-driven Method for Structural Optimization. https://doi.org/10.1016/B978-0-12-821330-8.00005-5

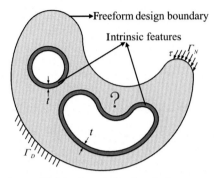

FIG. 5.1   Illustration of a specified freeform design domain for structural optimization.

auxiliary modeler of intrinsic features. Each submodeler is defined below and illustrated in Fig. 5.2

$$\begin{cases} \Omega_{FDDM}^{basic} = \Omega_{outer} \cap (\Omega_{inner})^C = (\Omega_1 \cup \Omega_2 \cup \Omega_3) \cap (\Omega_4 \cup \Omega_5 \cup \Omega_6 \cup \Omega_7)^C \\ \Omega_{FDDM}^{aux} = \Omega_8 \cup \Omega_9 \cup \Omega_{10} \cup \Omega_{11} \end{cases} \quad (5.1)$$

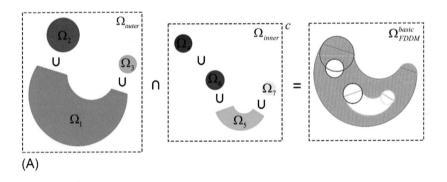

(A)

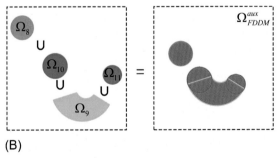

(B)

FIG. 5.2   Descriptions of the FDDM with two submodelers: (A) basic modeler of freeform design domain; and (B) auxiliary modeler of intrinsic features.

Consequently, the level-set functions of both submodelers are formulated as

$$
\begin{cases}
\Phi_{FDDM}^{basic}(\pmb{x},\pmb{p}) = \Phi_{outer} \wedge (-\Phi_{inner}) = (\Phi_1 \vee \Phi_2 \vee \Phi_3) \wedge (-(\Phi_4 \vee \Phi_5 \vee \Phi_6 \vee \Phi_7)) \\
\Phi_{FDDM}^{aux}(\pmb{x},\pmb{q}) = \Phi_8 \vee \Phi_9 \vee \Phi_{10} \vee \Phi_{11}
\end{cases}
\tag{5.2}
$$

Notice that the auxiliary modeler of intrinsic features is not mandatorily needed in the FDDM. In the above formulation, the negative sign represents the complement of a geometry. If the boundary shape of the freeform design domain and intrinsic features are designable, the corresponding shape parameters should be included in the set of design variables. In Eq. (5.2), $\pmb{p}$ and $\pmb{q}$ denote the set of design variables related to the basic and auxiliary modeler of FDDM, respectively.

Specially, suppose the auxiliary modeler of intrinsic features is attributed with uniform thickness t. Then, the level-set function Φ_{FDDM}^{aux} can be simplified when the inner domain Ω_{inner} is defined with a signed distance function.

$$
\Phi_{FDDM}^{aux}(\pmb{x},\pmb{q}) = \Phi_{inner} + t
\tag{5.3}
$$

Example 5.1 FDDM related to a torque arm.

The torque arm shown in Fig. 5.3A is considered a benchmark for shape optimization by many researchers (Zhang et al., 1995; Kim and Chang, 2005; Van Miegroet, 2012). In this example, there is no auxiliary modeler of intrinsic features. Both the inner and outer boundaries are supposed to be designable. A portion of the outer boundary is defined by a series of prescribed interpolation points $\{x_i, y_i\}$ $(i = 0,1, \ldots,N)$. $\pmb{Y} = \{y_0, y_1, \ldots, y_N\}$ belongs to the set of shape design variables while x_i $(i = 0,1, \ldots,N)$ are considered constants and sequenced in strict monotonic order. Based on the Boolean operations in Fig. 5.3B, Φ_{FDDM} is constructed by the following relation and is shown in Fig. 5.3C.

$$
\Phi_{FDDM}\left(x, y, \pmb{Y}, r_1, r_2, x_Q, x_R\right) = \Phi_1 \vee \Phi_3(\pmb{Y}) \wedge (-\Phi_2) \vee \Phi_4 \wedge (-\Phi_5) \wedge (-\Phi_{slot})
\tag{5.4}
$$

with

$$
\Phi_{slot} = \Phi_6\left(r_1, x_Q\right) \vee \Phi_7\left(r_1, r_2, x_Q, x_R\right) \vee \Phi_8(r_2, x_R)
\tag{5.5}
$$

where Φ_i $(i = 1,2,\cdots,8)$ denotes the level-set function of Ω_i in Fig. 5.3B. x_Q and x_R denote the x coordinates of circle centers Q and R marked in Fig. 5.3A.

5.1.2 Topology variation modeler

The TVM acts as a geometric tool whose functionality is to realize topology changes through hole creation, removal, and merging. Implicit representation in terms of level-set functions has the advantage of easily describing topology

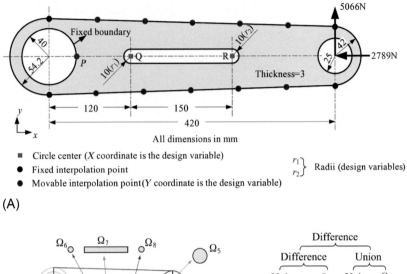

(A)

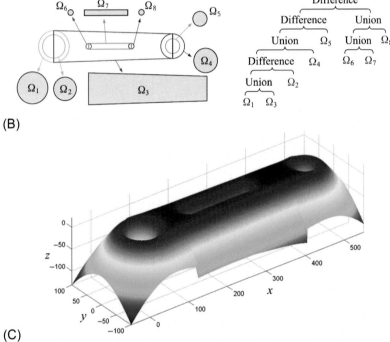

(B)

(C)

FIG. 5.3 The FDDM related to a torque arm: (A) the dimensions, boundary conditions, and design variables; (B) the constitutive primitives; and (C) the level-set function of the FDDM obtained by the R-function.

changes. As discussed below, two cases are considered for the construction of the level-set functions for TVM.

(i) *TVM constructed by basis functions*

Given an embedding domain D, the corresponding discrete level-set function could be constructed using the signed distance function to the boundary of Ω_{TVM}.

$$\Phi_{discrete}(x) = \begin{cases} \min_{x_p \in \partial\Omega_{TVM}} \|x - x_p\|, \ \forall x \in \Omega_{TVM} \\ -\min_{x_p \in \partial\Omega_{TVM}} \|x - x_p\|, \ \forall x \in D \setminus \Omega_{TVM} \end{cases} \quad (5.6)$$

The level-set function, Φ_{TVM}, can be constructed by interpolation of the basis functions to approximate the discrete level-set function. It corresponds to

$$\Phi_{TVM}(x, a) = \varphi^T(x)a = \sum_{i=1}^{n} \varphi_i(x)a_i \quad (5.7)$$

where n is the number of prescribed knots distributed in the embedding domain D. $\varphi(x) = (\varphi_1(x), \varphi_2(x), \ldots, \varphi_n(x))^T$ denotes the vector of basis functions and $a = (a_1, a_2, \ldots, a_n)^T$ denotes the vector of expansion coefficients adopted as the set of design variables. Here, basis functions can generally be selected according to B-splines (Chen et al., 2007), RBFs (Wang and Wang, 2006), CS-RBFs (Luo et al., 2008), etc.

Given $\Phi_{discrete}(x)$, the coefficient vector a can be obtained by solving the system of linear equations

$$\Phi_{discrete}(x) = \varphi^T(x)a \quad (5.8)$$

Example 5.2 TVM constructed by CS-RBFs in an L-shaped domain.
Suppose a number of circular holes in Fig. 5.4A are prescribed within the L-shaped design domain as the initial topology. The corresponding signed distance function defined by Eq. (5.6) is interpolated by CS-RBFs in Section 2.3.2 to offer a freeform implicit representation of hole distributions in the L-shaped beam and is plotted in Fig. 5.4B.

In order to interpolate the signed distance function in Fig. 5.4B by CS-RBFs, 80×80 knots are uniformly distributed within a 1×1 square domain shown in Fig. 5.4C. The support radius involved in the CS-RBFs in Eq. (2.47) is assumed to be $d_p = 0.03$. Moreover, the knots are classified into active ones and inactive ones that are located outside the L-shaped domain. The distance from an arbitrary inactive knot to the design domain's boundary is greater than the support radius d_p. As the CS-RBFs centered at the inactive knots have no contribution to the design domain, the corresponding expansion coefficients a_i are not considered as design variables. The constructed level-set function $\Phi_{TVM}(x, a)$ is depicted in Fig. 5.4D.

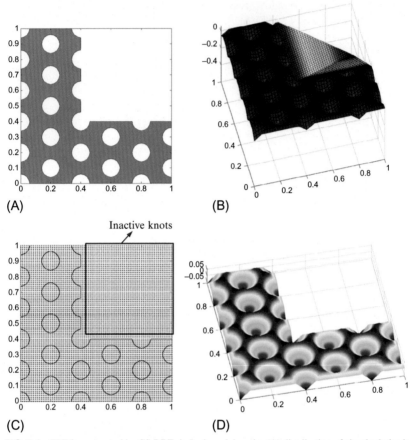

FIG. 5.4   TVM constructed by CS-RBFs in L-shaped domain: (A) distribution of circular holes in L-shaped domain; (B) the corresponding signed distance function; (C) distribution and classification of 80×80 knots; and (D) level-set function Φ_{TVM} excluding the portion outside the design domain.

(ii) *TVM constructed with solid and void features via Boolean operations*

As illustrated in Fig. 5.5C, suppose there exist n_1 solid features Ω_1^s, Ω_2^s, ..., Ω_{n1}^s and n_2 void features Ω_1^v, Ω_2^v, ..., Ω_{n2}^v in the embedding domain D with their level-set functions satisfying

$$\Phi_i^s(\boldsymbol{x}) \begin{cases} > 0, & \text{in } \Omega_i^s \\ = 0, & \text{on } \partial\Omega_i^s \\ < 0, & \text{in } D \backslash \bar{\Omega}_i^s \end{cases} \tag{5.9}$$

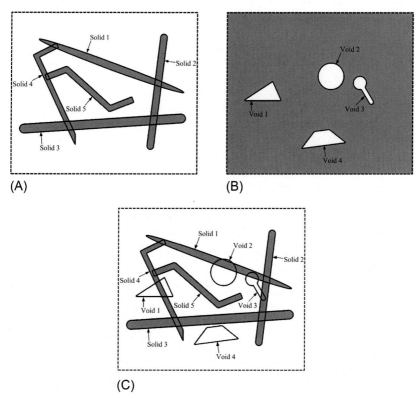

FIG. 5.5 Feature definition with solid and (or) void features: (A) solid features; (B) void features; and (C) mixed solid-void features.

and

$$\Phi_j^v(x) \begin{cases} < 0, & \text{in } \Omega_j^v \\ = 0, & \text{on } \partial\Omega_j^v \\ > 0, & \text{in } D \backslash \bar{\Omega}_j^v \end{cases} \tag{5.10}$$

The TVM can hence be constructed with mixed solid-void features by means of Boolean operations as

$$\Omega_{TVM} = \Omega^s \cap \Omega^v = \left(\Omega_1^s \cup \Omega_2^s \cup \cdots \cup \Omega_{n_1}^s \right) \cap \left(\Omega_1^v \cap \Omega_2^v \cap \cdots \cap \Omega_{n_2}^v \right) \tag{5.11}$$

In this sense, the TVMs defined by pure solid or pure void features are special cases when the number of void features n_2 or the number of solid features n_1 equals zero, as illustrated in Figs. 5.5A and B.

Correspondingly, the level-set functions defined with pure solid, pure void, and mixed solid-void features can be written as

(a) Level-set function of TVM defined with pure solid features

$$\Phi_{TVM}^s(x, s_1, s_2, \cdots, s_{n_1}) = \Phi_1^s \vee \Phi_2^s \vee \cdots \vee \Phi_{n_1}^s \qquad (5.12)$$

(b) Level-set function of TVM defined with pure void features

$$\Phi_{TVM}^v(x, v_1, v_2, \cdots, v_{n_2}) = \Phi_1^v \wedge \Phi_2^v \wedge \cdots \wedge \Phi_{n_2}^v \qquad (5.13)$$

(c) Level-set function of TVM defined with mixed solid-void features

$$\Phi_{TVM}(x, s_1, s_2, \cdots, s_{n_1}, v_1, v_2, \cdots, v_{n_2}) = \Phi_{TVM}^s \wedge \Phi_{TVM}^v \qquad (5.14)$$

Here, the vectors of design variables s_i ($i = 1,2,\cdots, n_1$) and v_j ($j = 1,2,\cdots, n_2$) include position coordinates and the inclined angle and shape variables of the ith solid feature and the jth void feature, respectively. The resulting level-set functions of the above TVMs are shown in Fig. 5.6.

Example 5.3 TVM constructed with CBSs as freeform features.
The CBS introduced in Section 2.3.3 can be regarded as a freeform feature with a certain degree of deformation ability according to the number of control radii. In this example, CBSs are used as basic primitives to construct the geometric model of a feature-based TVM. Fig. 5.7A shows a set of CBSs defined in the embedding domain D. Each CBS represents a hole instead of a solid inclusion. The level-set function Φ_{TVM} is thus stated as

$$\Phi_{TVM}(x, v) = \overset{n_2}{\underset{i=1}{\wedge}} (-\Phi_i(x, v_i)) \qquad (5.15)$$

where $\Phi_i(x, v_i)$ is the level-set function of the ith CBS as described in Eq. (2.58). v is the vector of design variables consisting of the center positions and control radii of all CBSs. The resulting higher-dimensional level-set function of the TVM defined with holed CBS is shown in Fig. 5.7B.

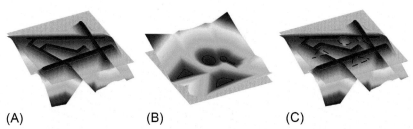

(A)   (B)   (C)

FIG. 5.6 The resulting level-set functions of TVMs defined with solid and (or) void features: (A) solid features; (B) void features; and (C) mixed solid-void features.

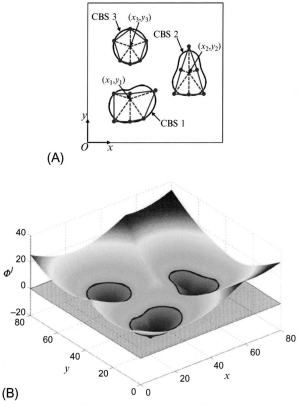

FIG. 5.7   TVM defined with freeform features of CBSs: (A) a set of CBSs in embedding domain D; and (B) the resulting level-set function of TVM.

5.1.3   Action of the TVM onto the FDDM

As presented above, both modelers hold their own functionalities. The topology optimization of a structure is interpreted as the application process of the TVM onto the FDDM. The traditional rectangular or even the L-shaped design domains in topology optimization can thus be regarded as simplified cases.

The application process consists of two steps. First, the TVM is combined with the auxiliary modeler of intrinsic features through Boolean union. Then, the basic modeler of FDDM is further clipped via Boolean intersection. In other words, the FDDM acts as a working window so that only its inner part constitutes the effective part of a structure. The integrated design model hence corresponds to

$$\Omega = \Omega_{TVM} \cup \Omega_{FDDM}^{aux} \cap \Omega_{FDDM}^{basic} \qquad (5.16)$$

Correspondingly, its level-set function Φ can be expressed as

$$\Phi(x,d) = \Phi_{TVM} \vee \Phi_{FDDM}^{aux} \wedge \Phi_{FDDM}^{basic} \qquad (5.17)$$

In Eq. (5.17), the involved parameters in both TVM and FDDM define the set of design variables d. These parameters can refer to expansion coefficients a of basis functions in Eq. (5.7) or featured design variables $\{s_i, v_j\}$ of solid-void features in Eqs. (5.12)–(5.14) defining the TVM and geometric parameters $\{p, q\}$ of primitives such as sizes, radii, locations, and angles in the FDDM. To make things clear, Eq. (5.17) is highlighted by considering the design of a bracket structure. Two design cases are studied below.

- Design case 1: Fig. 5.8A illustrates a freeform design domain of the bracket limited by curved boundaries for topology optimization.
- Design case 2: Fig. 5.8B illustrates four holes with a fixed width prescribed as the nondesignable solid features, that is, intrinsic features in the topology optimization process.

5.2   Problem statement of feature-driven optimization

5.2.1   Mathematical formulations

In this book, the structural optimization problem of linear elasticity is considered. The mathematical formulation is generally stated as

$$
\begin{aligned}
&\text{Find}: \quad d \\
&\text{Min}: \quad \Pi(d) \\
&\text{s.t.} \quad \begin{cases} g_j(d) \leq \overline{g}_j & j = 1, 2, \ldots, m \\ \underline{d} \leq d \leq \overline{d} \end{cases}
\end{aligned}
\qquad (5.18)
$$

where d is the set of design variables bounded within interval $\left[\underline{d}, \overline{d}\right]$. The aim is to find the optimal value of d for the structure with the minimum value of objective function $\Pi(d)$ under m prescribed constraints g_j ($j = 1, 2, \ldots, m$). Notice that the case of maximizing an objective function $\Pi(d)$ can equivalently be converted into minimizing the negative of the objective function, that is, $-\Pi(d)$.

As is known, the structural weight and mechanical performance (e.g., structural stiffness and maximum stress level) are critical for a structure subjected to external loading. In the following, several formulations widely studied in the context of structural optimization are presented.

The first one concerns the least-weight design of a structure with stress constraints. It can mathematically be stated as

$$
\begin{aligned}
&\text{Find}: \quad d \\
&\text{Min}: \quad V(\Phi) = \int_D H(\Phi(x,d))d\Omega \\
&\text{s.t.} \quad \begin{cases} \sigma_j(u) \leq \overline{\sigma} & j = 1, 2, \ldots, m \\ \underline{d} \leq d \leq \overline{d} \end{cases}
\end{aligned}
\qquad (5.19)
$$

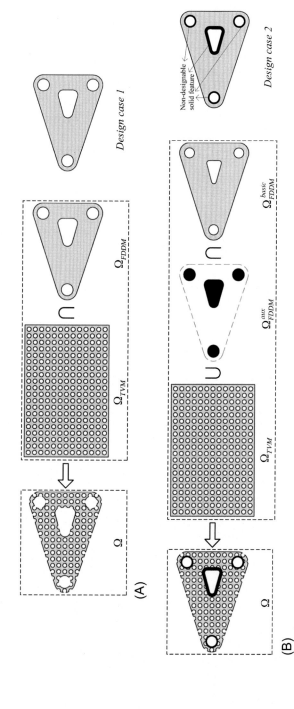

FIG. 5.8 Boolean operations between FDDM and TVM interpolated by basis function: (A) design case 1; and (B) design case 2.

in which $V(\Phi)$ denotes the volume of the structure in terms of the level-set function Φ. σ_j ($j=1,2,\cdots,m$) denotes the representative stress measures such as von Mises stresses on sampling points of number m that are large enough to provide a consistent description of the stress level. $\bar{\sigma}$ is the upper bound of the stress constraints.

The second formulation concerns the minimum mean compliance design of a structure with a prescribed material usage. It can mathematically be stated as

Find : $\boldsymbol{d}$

Min : $J = \dfrac{1}{2}\displaystyle\int_D E_{ijkl}\varepsilon_{ij}(\boldsymbol{u})\varepsilon_{kl}(\boldsymbol{u})H(\Phi(\boldsymbol{x},\boldsymbol{d}))d\Omega = \dfrac{1}{2}\boldsymbol{U}^T\boldsymbol{K}\boldsymbol{U}$

$$\text{s.t.} \quad \begin{cases} V(\boldsymbol{d}) = \displaystyle\int_D H(\Phi(\boldsymbol{x},\boldsymbol{d}))d\Omega \leq \overline{V} \\ \underline{\boldsymbol{d}} \leq \boldsymbol{d} \leq \overline{\boldsymbol{d}} \end{cases}$$

(5.20)

in which J denotes the structural mean compliance. $\overline{V}$ refers to the upper bound of the prescribed volume constraint.

The third formulation concerns the minimization of the mean compliance of a structure subjected to both volume and stress constraints. It can mathematically be stated as

Find : $\boldsymbol{d}$

Min : $J = \dfrac{1}{2}\displaystyle\int_D E_{ijkl}\varepsilon_{ij}(\boldsymbol{u})\varepsilon_{kl}(\boldsymbol{u})H(\Phi(\boldsymbol{x},\boldsymbol{d}))d\Omega = \dfrac{1}{2}\boldsymbol{U}^T\boldsymbol{K}\boldsymbol{U}$

$$\text{s.t.} \quad \begin{cases} V(\boldsymbol{d}) = \displaystyle\int_D H(\Phi(\boldsymbol{x},\boldsymbol{d}))d\Omega \leq \overline{V} \\ \sigma_j(\boldsymbol{d}) \leq \bar{\sigma} \quad j=1,2,\ldots,m \\ \underline{\boldsymbol{d}} \leq \boldsymbol{d} \leq \overline{\boldsymbol{d}} \end{cases}$$

(5.21)

5.2.2 Numerical treatments of active stress constraints

Due to the presence of a large number of stress constraints in the formulations of Eqs. (5.19) and (5.21), numerical treatments are hence needed to deal with the stress constraints for the improvement of optimization efficiency. In this section, the numerical method of dynamic aggregation is presented. It is used to cluster the active stress constraints into a small number of aggregated constraints without much sacrificing the precision of the local stress level control. The aggregation technique can be carried out with the help of envelope functions such as the Ricci function and the KS function. Take the KS function as an example. The aggregated stress value is stated as

$$\sigma_{KS} = \frac{1}{\eta}\ln\left[\sum_{j=1}^{m} e^{\eta \cdot \sigma_j}\right], \quad \eta > 0 \tag{5.22}$$

It is bounded by the following relation (Luo et al., 2013)

$$\sigma_{max} \leq \sigma_{KS} \leq \sigma_{max} + \frac{\ln(m)}{\eta} \tag{5.23}$$

with $\sigma_{max} = \max_{j=1,2,\ \ldots,m}\{\sigma_j\}$.

The dynamic aggregation technique consists of gradually increasing the aggregation parameter η that dominates the approximation quality of the KS function or Ricci function during the course of optimization (James et al., 2009; Lee et al., 2012). As the set of local stress constraints in Eqs. (5.19) and (5.21) is equivalent to $\sigma_{max} \leq \bar{\sigma}$, the global approximation $\sigma_{KS} \leq \bar{\sigma}$ is conservative according to Eq. (5.23) and the approximation error could be reduced by narrowing the gap between σ_{max} and σ_{KS}. To do this, two ways exist: one is to decrease the number of constraints, m, by means of the active-set strategy and the grouped aggregation; the other is to increase the value of the aggregation parameter η dynamically.

With this idea in mind, local stress constraints satisfying $\sigma_j \geq \beta\bar{\sigma}$ ($0<\beta<1$) are first selected as the current set of potentially active constraints. They are further divided into a relatively small number of groups, and each group is finally aggregated into one constraint by the KS function with a dynamic aggregation parameter η. The division method used here is the same as the interlacing regional stress measure strategy in Le et al. (2010). At a certain iteration, the stress values related to the active selection are sorted in descending order, as $\sigma_1 \geq \sigma_2 \geq \ldots \geq \sigma_{ma}$ (m_a is the number of selected active constraints) and are further divided into M groups

$$G_k = \{\sigma_k, \sigma_{k+M}, \sigma_{k+2M}, \cdots\sigma_{k+(m_a-M)}\} \quad k=1,2,\ldots,M \tag{5.24}$$

where G_k is the set of stress values related to the kth group, and M is the total number of groups, which remains unchanged during the whole optimization process. It should be noted that the number m_a in every iteration must be an integral multiple of M. So, some of the most critical constraints in the currently inactive set should be added into the active selection if m_a is not divisible by M.

Before the grouped aggregation operation, a normalization process is carried out to divide the stress values by the stress limit $\bar{\sigma}$, so that local stress constraints in Eqs. (5.19) and (5.21) are finally replaced by

$$\sigma_{G_k} = \frac{1}{\eta} \ln\left[\sum_{i=1}^{m_a/M} e^{\eta\frac{\sigma_{k+(i-1)M}}{\bar{\sigma}}}\right] \leq 1 \quad k=1,2,\ldots,M \tag{5.25}$$

Thus, m_a/M, the number of local constraints clustered by each aggregation function in Eq. (5.25), is much smaller than m in Eqs. (5.19) and (5.21). It can further be reduced because the number of potentially active constraints selected decreases along with the optimization process. In fact, there are two main merits benefiting from the dynamically increased parameter η: One is the prevention of the iteration from premature convergence to local minima to a certain degree.

Because small values of η are adopted at the beginning, the feasible region of Eq. (5.25) becomes very conservative and is not dominated only by the most violated constraints that change frequently. The other is the achievement of a good approximation of the feasible region at the subsequent stage of optimization because of the successive increase of parameter η.

5.3 Feature-based sensitivity analysis

Sensitivity analysis is essential for gradient-based optimization algorithms. For a linear elastic structure, the key to sensitivity analysis is to calculate the sensitivities of the load vector and the stiffness matrix with respect to design variables. Also, the sensitivities of the aggregated stress constraints in Eq. (5.25) could be derived by using the sensitivity of the displacement vector $\partial U/\partial d_i$ and the chain rule of differentiation.

First, the sensitivity of J with respect to the ith design variable d_i is calculated as

$$\frac{\partial J}{\partial d_i} = U^T K \frac{\partial U}{\partial d_i} + \frac{1}{2} U^T \frac{\partial K}{\partial d_i} U \tag{5.26}$$

In this chapter and Chapter 6, the load vector F is supposed to be design-independent (i.e., $\partial F/\partial d_i = 0$). More complex cases including design-dependent loads will be discussed in Chapter 7. Therefore, it follows that

$$\frac{\partial U}{\partial d_i} = -K^{-1} \frac{\partial K}{\partial d_i} U \tag{5.27}$$

The substitution into Eq. (5.26) results in

$$\frac{\partial J}{\partial d_i} = -\frac{1}{2} U^T \frac{\partial K}{\partial d_i} U \tag{5.28}$$

Second, let us consider the sensitivity analysis of stress constraints. For a 2D problem, the von Mises stress of the jth material point is defined as

$$\sigma_{von}(x_j) = \sqrt{\frac{\sigma_x^2(x_j) + \sigma_y^2(x_j) + (\sigma_x(x_j) - \sigma_y(x_j))^2 + 6\tau_{xy}^2(x_j)}{2}}$$

$$= \sqrt{\frac{\tilde{\sigma}^T(x_j)\tilde{\sigma}(x_j)}{2}} \tag{5.29}$$

with

$$\tilde{\sigma}(x_j) = \left\{ \sigma_x(x_j), \sigma_y(x_j), \sigma_x(x_j) - \sigma_y(x_j), \sqrt{6}\tau_{xy}(x_j) \right\}^T \tag{5.30}$$

where $\sigma_x(x_j)$, $\sigma_y(x_j)$, and $\tau_{xy}(x_j)$ denote three stress components. Then, the sensitivity of $\sigma_{von}(x_j)$ with respect to the ith design variable d_i is calculated as

$$\frac{\partial \sigma_{von}(x_j)}{\partial d_i} = \frac{1}{2\sigma_{von}(x_j)} \tilde{\sigma}^T(x_j) \frac{\partial \tilde{\sigma}(x_j)}{\partial d_i} \qquad (5.31)$$

The sensitivity of $\sigma(x_j) = \{\sigma_x(x_j), \sigma_y(x_j), \tau_{xy}(x_j)\}^T$ is computed as

$$\frac{\partial \sigma(x_j)}{\partial d_i} = DB(x_j) \frac{\partial U}{\partial d_i} \qquad (5.32)$$

where $\partial U / \partial d_i$ can be obtained by Eq. (5.27).

Clearly, the basic computing is the sensitivity of stiffness matrices. To do this, consider the general case of a functional integral

$$\Psi = \int_D f(x) H(\Phi(x, d)) d\Omega \qquad (5.33)$$

To be specific, Ψ corresponds to the structural stiffness matrix when $f(x) = B^T D B$ and represents structural volume when $f(x) = 1$. To calculate the sensitivities of Ψ, two sensitivity analysis schemes, i.e., the domain integral and the boundary integral, have been developed within the framework of fixed mesh. The introductions and comparisons of both schemes are given below.

5.3.1 Sensitivity analysis with the domain integral scheme

According to the chain rule, the sensitivity of Ψ can be stated as

$$\frac{\partial \Psi(\Phi)}{\partial d_i} = \int_D f(x) \frac{\partial H(\Phi)}{\partial \Phi} \frac{\partial \Phi}{\partial d_i} d\Omega = \int_D f(x) \delta(\Phi) \frac{\partial \Phi}{\partial d_i} d\Omega \qquad (5.34)$$

where $\delta(\Phi)$ is the Dirac delta function referring to the derivative of the Heaviside function, which is expressed as

$$\delta(\Phi) = \frac{\partial H(\Phi)}{\partial \Phi} = \begin{cases} +\infty, & \Phi = 0 \\ 0, & \Phi \neq 0 \end{cases} \qquad (5.35)$$

It is well known that the singularity of $\delta(\Phi)$ usually causes difficulties in computing the sensitivity in Eq. (5.34). To circumvent this difficulty, one feasible way is to use the regularized Heaviside function $\hat{H}(\Phi)$ mentioned in Section 2.2.2 to replace the discontinuous Heaviside function $H(\Phi)$. In this sense, the Dirac delta function $\delta(\Phi)$ is regularized to make the computing of sensitivity in Eq. (5.34) possible. To give an example, the regularized Heaviside function in the form of a piecewise polynomial is stated as

$$\hat{H}(\Phi) = \begin{cases} 1, & \Phi \geq \Delta \\ \dfrac{3(1-\alpha)}{4}\left(\dfrac{\Phi}{\Delta} - \dfrac{\Phi^3}{3\Delta^3}\right) + \dfrac{1+\alpha}{2}, & -\Delta \leq \Phi < \Delta \\ \alpha, & \Phi < -\Delta \end{cases} \qquad (5.36)$$

Its derivative (i.e., the regularized Dirac delta function) reads

$$\hat{\delta}(\varPhi) = \begin{cases} \dfrac{3(1-\alpha)}{4}\left(\dfrac{1}{\Delta}-\dfrac{\varPhi^2}{\Delta^3}\right), & -\Delta \le \varPhi < \Delta \\ 0, & \text{elsewhere} \end{cases} \tag{5.37}$$

in which $\alpha = 10^{-3}-10^{-12}$ is a small value of a fictitious material in the replacement of the void material to prevent the ill-conditioning of the stiffness matrix.

Obviously, the regularized $\hat{\delta}(\varPhi)$ is nonzero only in a narrow-band domain around the structural boundary where $-\Delta \le \varPhi < \Delta$. Thus, the narrow-band domain integral scheme for sensitivity analysis is then developed as

$$\frac{\partial \varPsi(\varPhi)}{\partial d_i} = \int_D f(x)\frac{\partial \hat{H}(\varPhi)}{\partial \varPhi}\frac{\partial \varPhi}{\partial d_i}d\Omega = \int_D f(x)\hat{\delta}(\varPhi)\frac{\partial \varPhi}{\partial d_i}d\Omega \tag{5.38}$$

Within the framework of FCM, the above sensitivities are calculated in the narrow-band domain within the boundary cells Ω_c. For each boundary cell Ω_c, the narrow-band domain is determined by means of the boundary contour and the value of Δ. The adaptive quadtree technique can be used to enrich the Gauss points inside the narrow-band domain for computing. In this way, boundary variations even with large curvatures and shape corners can be well captured by the adaptive refinement and preserved in sensitivity analysis.

Specially, the sensitivities of the cell stiffness matrix and the cell volume hence read

$$\begin{cases} \dfrac{\partial k_c}{\partial d_i} = \displaystyle\int_{\Omega_c} \boldsymbol{B}^T\boldsymbol{D}\boldsymbol{B}\hat{\delta}(\varPhi)\dfrac{\partial \varPhi}{\partial d_i}d\Omega \\ \dfrac{\partial v_c}{\partial d_i} = \displaystyle\int_{\Omega_c} \hat{\delta}(\varPhi)\dfrac{\partial \varPhi}{\partial d_i}d\Omega \end{cases} \tag{5.39}$$

Obviously, the mathematical expressions of the level-set function $\varPhi$, the value of Δ, and the size of the elements are critical for sensitivity analysis accuracy and the distribution of gray materials. A numerical example is given here to show the effects of related parameters.

Example 5.4 Effects of narrow-band width on structural analysis accuracy. Take the plate with a hole in Fig. 5.9 as an example. The left side of the plate is fixed and a vertical force is applied at the lower right corner. The structure has a dimension of 20×20 and the radius of the circular hole centered at $(10, 10)$ is 5. Thus, the level-set function of the whole structure corresponds to

$$\varPhi = \sqrt{(x-10)^2 + (y-10)^2} - 5 \tag{5.40}$$

The first step is to embed the structure into a regular domain with a dimension of 20×20. The domain is further discretized into rectangular elements of size 0.5×0.5. In this example, the value of the level-set function at the center

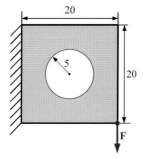

FIG. 5.9 A plate with a circular hole.

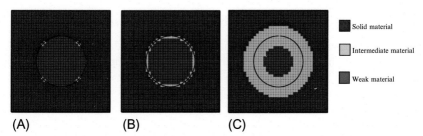

Solid material

Intermediate material

Weak material

(A)   (B)   (C)

FIG. 5.10   Material distributions of the plate with inner circular hole: (A) $\Delta = 0.05$; (B) $\Delta = 0.25$; and (C) $\Delta = 2$.

point of each cell is utilized for the assignment of material property. Fig. 5.10 gives the material distributions with three different values of Δ. Obviously, the number of elements full of intermediate materials increases when Δ is set to a larger value. At the same time, the sensitivity accuracy cannot be guaranteed when Δ is too small.

The effects of Δ upon compliance and volume are further investigated and illustrated in Fig. 5.11A and B, respectively. When $\Delta = 0.24$, the compliance has the same value as the solution of finite element analysis. With the increase of Δ, the compliance keeps coming down and considerable errors are produced. The volume also has a downward trend except for a small segment at the beginning and reaches the analytic solution at $\Delta = 0.27$. In view of this, nearly half the element size seems to be a good choice for Δ.

Notice that when the narrow-band scheme is used, the level-set function should be normalized into a quasiequidistant isocontour in sensitivity analysis by means of the first-order approximation of signed distance function. This treatment is important for the attainment of a clear material distribution in topology optimization.

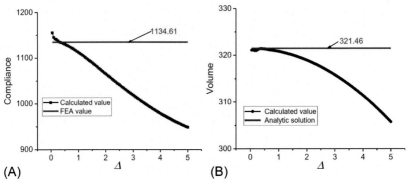

FIG. 5.11   The effect of Δ selection on compliance and volume: (A) compliance; and (B) volume.

5.3.2 Sensitivity analysis with boundary integral scheme

An alternative scheme to calculate the sensitivity in Eq. (5.34) can be obtained by introducing the directional derivative of the Heaviside function in the normal direction

$$
\begin{aligned}
\widetilde{\delta}(x) &= \nabla H(\varPhi(x)) \cdot n \\
&= \frac{\partial H(\varPhi(x))}{\partial \varPhi} \nabla \varPhi(x) \cdot \frac{\nabla \varPhi(x)}{\|\nabla \varPhi(x)\|} \\
&= \frac{\partial H(\varPhi(x))}{\partial \varPhi} \|\nabla \varPhi(x)\|
\end{aligned}
\tag{5.41}
$$

Eq. (5.34) can thus be rewritten in the form of a boundary integral as

$$
\begin{aligned}
\frac{\partial \varPsi(\varPhi)}{\partial d_i} &= \int_D f(x) \frac{\partial H(\varPhi)}{\partial \varPhi} \frac{\partial \varPhi}{\partial d_i} d\Omega \\
&= \int_D f(x) \frac{\partial \varPhi}{\partial d_i} \frac{1}{\|\nabla \varPhi\|} \left(\frac{\partial H(\varPhi)}{\partial \varPhi} \|\nabla \varPhi\| \right) d\Omega \\
&= \int_D f(x) \frac{\partial \varPhi}{\partial d_i} \frac{1}{\|\nabla \varPhi\|} \widetilde{\delta}(x) d\Omega \\
&= \int_{\partial \Omega} f(x) \frac{\partial \varPhi}{\partial d_i} \frac{1}{\|\nabla \varPhi\|} d\Gamma
\end{aligned}
\tag{5.42}
$$

Likewise, the sensitivities of the cell stiffness matrix and the cell volume correspond to

$$
\begin{cases}
\dfrac{\partial k_c}{\partial d_i} = \displaystyle\int_{\partial \Omega_c} B^T D B \frac{\partial \varPhi}{\partial d_i} \frac{1}{\|\nabla \varPhi\|} d\Gamma \\[2ex]
\dfrac{\partial v_c}{\partial d_i} = \displaystyle\int_{\partial \Omega_c} \frac{\partial \varPhi}{\partial d_i} \frac{1}{\|\nabla \varPhi\|} d\Gamma
\end{cases}
\tag{5.43}
$$

where $\partial\Omega_c$ refers to the cth boundary cell interface belonging to a part of the structural boundary. The above relation means that derivative computing can only be done for boundary cells.

Notice that fictitious parts of boundary cells might be taken into account by introducing the small value $\alpha = 10^{-3}$–10^{-12} in the integral computation. As a result, the stiffness matrix of the cth boundary cell is corrected as

$$k_c = \int_{\Omega_c} B^T DBH(\Phi(x))d\Omega + \int_{\Omega_c} \alpha B^T DB(1 - H(\Phi(x)))d\Omega \qquad (5.44)$$

And its sensitivity is calculated as

$$\frac{\partial k_c}{\partial d_i} = (1-\alpha)\int_{\partial\Omega_c} B^T DB \frac{\partial\Phi(x)}{\partial\alpha_i} \frac{1}{\|\nabla\Phi(x)\|}d\Gamma \qquad (5.45)$$

Numerical experience indicates that the sensitivity computed by Eq. (5.45) is almost the same as that computed by Eq. (5.43).

Here, we intend to emphasize the independence between the accuracy of the boundary integral scheme and the mathematical expressions of the level-set function Φ. By differentiating $\Phi(x, d) = 0$ on both sides using the chain rule, we can obtain the so-called Hamilton-Jacobi equation in terms of design variable d_i

$$\frac{\partial\Phi}{\partial d_i} + \nabla\Phi^T \frac{\partial x}{\partial d_i} = 0 \qquad (5.46)$$

Hence, the following relationship holds

$$\frac{\partial\Phi}{\partial d_i} \frac{1}{\|\nabla\Phi\|} = -\left(\frac{\nabla\Phi}{\|\nabla\Phi\|}\right)^T \frac{\partial x}{\partial d_i} = -n^T \frac{\partial x}{\partial d_i} \qquad (5.47)$$

Due to the fact that n and $\partial x/\partial d_i$ on the structural boundary are identical for different level-set functions, it is manifest that the sensitivity of the boundary integral scheme calculated on the actual structural boundary is independent of the mathematical expression of the level-set function.

Detailed comparisons of both schemes are listed in Table 5.1.

Example 5.5 Effect of the FCM order on sensitivity accuracy with the boundary integral scheme.
Consider a hollow cylindrical disk loaded uniformly by an outer pressure, as illustrated in Fig. 5.12A. This classical plane stress problem is adopted here to verify the accuracy of sensitivity analysis formulated in this section when the B-spline FCM is used in structural optimization.

As is well known, the solution of this problem corresponds to Lamé stress

$$\begin{cases} \sigma_r = -\dfrac{Pa^2}{a^2 - b^2}\left(1 - \dfrac{b^2}{r^2}\right) \\[2mm] \sigma_\theta = -\dfrac{Pa^2}{a^2 - b^2}\left(1 + \dfrac{b^2}{r^2}\right) \\[2mm] \tau_{r\theta} = 0 \end{cases} \qquad (5.48)$$

TABLE 5.1 Comparisons of boundary integral and domain integral schemes.

Scheme	Boundary integral	Domain integral
Mathematical formulation	$\dfrac{\partial \Psi(\Phi)}{\partial d_i} = \displaystyle\int_{\partial\Omega_c} f(\boldsymbol{x})\dfrac{\partial \Phi}{\partial d_i}\dfrac{1}{\|\nabla\Phi\|}d\Gamma$	$\dfrac{\partial \Psi(\Phi)}{\partial d} = \displaystyle\int_{\Omega_c} f(\boldsymbol{x})\hat{\delta}(\Phi)\dfrac{\partial \Phi}{\partial d}d\Omega$
Schematic		
Detail	a. The bisection method is used to find the end points of the structural boundary within a boundary cell b. The structural boundary is approximated by a straight line segment connecting two end points for the selection of Gauss points within a boundary cell	a. Boundary cells are partitioned with a quadtree refinement scheme b. Gauss points located inside the narrow band domain whose level-set function Φ takes value in $[-\Delta, \Delta]$ are used for sensitivity calculation
Notation	Independence of the mathematical expressions of the level-set function Φ	Signed distance function is needed to avoid blocks of gray regions

in which σ_r, σ_θ, and $\tau_{r\theta}$ are the radial, circumferential (hoop), and shear stresses. The definitions of other symbols can refer to Fig. 5.12A. Suppose the inner radius b is a design variable. The sensitivities of σ_r and σ_θ are then calculated as

$$\begin{cases} \dfrac{\partial \sigma_r}{\partial b} = \dfrac{2Pa^2 b(a^2 - r^2)}{r^2(a^2 - b^2)^2} \\[3mm] \dfrac{\partial \sigma_\theta}{\partial b} = \dfrac{2Pa^2 b(a^2 + r^2)}{r^2(a^2 - b^2)^2} \end{cases} \tag{5.49}$$

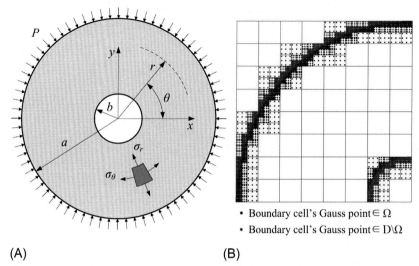

- Boundary cell's Gauss point $\in \Omega$
- Boundary cell's Gauss point $\in D\backslash\Omega$

(A)   (B)

FIG. 5.12   A hollow cylindrical disk under outer pressure: (A) cross-section of the cylinder; and (B) the FCM mesh for a quarter of the cross-section and Gauss points of boundary cells.

Due to the symmetry of the structural geometry and boundary conditions, only one quarter of the structure is considered here. It is embedded in a square domain discretized with square cells of the B-spline FCM, as illustrated in Fig. 5.12B. Relevant parameters are $a = 4$, $b = 1$, and $P = 1$. The adaptive integration is used to obtain high precision results for boundary cells. Fig. 5.12B shows the aggregation of Gauss points around the boundary that cuts through the cells.

In Fig. 5.13, the numerical sensitivity results of $\partial\sigma_r/\partial b$ and $\partial\sigma_\theta/\partial b$ transformed from the Cartesian coordinate system are compared with the corresponding analytical results of Eq. (5.49). Clearly, a good consistence exists only in the case of the bi-cubic B-spline cells ($p = q = 3$). The inconsistence with bi-quadratic ones ($p = q = 2$) is due to the fact that the number of 16×16 cells is not enough to ensure computing accuracy. The discontinuity of stress sensitivity occurs for bi-linear cells ($p = q = 1$) due to the C^0-continuity of basis functions along the cell interfaces, as indicated in Fig. 5.14. Therefore, $p \geq 2$ and $q \geq 2$ should be used at least for B-spline basis functions.

To have a clear idea, the relative error of numerical sensitivity solution in the L^2 norm, e_{L2}, is defined to measure the discretization error.

$$e_{L2} = \sqrt{\frac{\int_D \left(S_\sigma^* - S_\sigma\right)^T \left(S_\sigma^* - S_\sigma\right) H(\Phi(x,b)) d\Omega}{\int_D \left(S_\sigma^*\right)^T \left(S_\sigma^*\right) H(\Phi(x,b)) d\Omega}} \qquad (5.50)$$

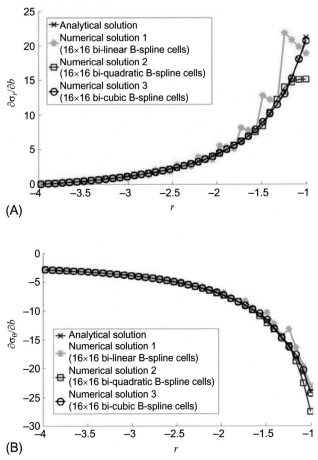

FIG. 5.13 Comparisons between the analytical and numerical sensitivity results along the radial direction: (A) radial stress sensitivities $\partial \sigma_r / \partial b$; and (B) circumferential stress sensitivities $\partial \sigma_\theta / \partial b$.

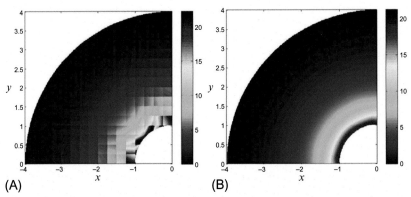

FIG. 5.14 Numerical sensitivity contours of radial stress $\partial \sigma_r / \partial b$ in two different interpolation cases: (A) 16×16 bi-linear B-spline cells with $p = q = 1$; and (B) 16×16 bi-cubic B-spline cells with $p = q = 3$.

in which S_σ^* is the exact stress sensitivity written as

$$S_\sigma^* = \left\{ \frac{\partial \sigma_r}{\partial b}, \frac{\partial \sigma_\theta}{\partial b}, \frac{\partial \tau_{r\theta}}{\partial b} \right\}^T \qquad (5.51)$$

The first two elements of S_σ^* are calculated by Eq. (5.49) while the last term is zero in our case. S_σ is the corresponding B-spline FCM solution. The convergence is shown in Fig. 5.15A for e_{L2} of the L^2 norm versus h-refinement. The relationship between e_{L2} and the computing time t is also shown in Fig. 5.15B.

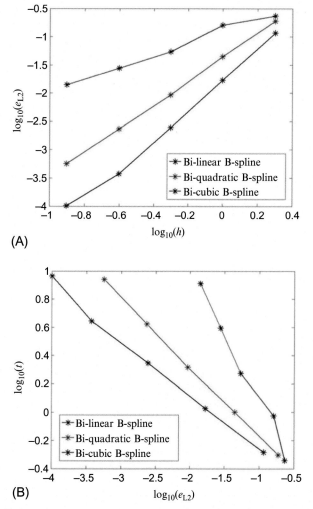

(A)

(B)

FIG. 5.15   Performance comparisons of B-spline basis functions of different orders for stress sensitivity: (A) relationship between e_{L2} and h-refinement; and (B) relationship between computing time t and e_{L2}.

It can be seen that the curve of the bi-cubic B-spline is always beneath the other two ones. As a result, the computing time of the bi-cubic B-spline is the shortest for the same sensitivity accuracy. However, it should be noted that an increase of the B-spline's order could not always reduce the computing time because high-order basis functions make the stiffness matrix very dense. Therefore, orders and continuities should properly be balanced for the B-spline FCM depending upon the nature of specific problems.

5.3.3 Sensitivity property with design domain preserving

It is important to note that the violation of the design domain boundary might happen along with the zero level-set movements in existing level-set methods (Rong and Liang, 2008). As shown in this section, this problem is completely avoided by the current formulation and its sensitivity property.

Discussions are made separately according to the classification of the aforementioned two design cases with or without nondesignable solid features in Section 5.1.3.

● Design case 1: Freeform design domain without nondesignable solid features

Suppose the TVM related to Ω_{TVM} is constructed by the CS-RBFs while the FDDM related to Ω_{FDDM} is constructed by the R-function. Each is represented by the level-set function $\Phi_{TVM}(x,a)$ and $\Phi_{FDDM}(x,p)$, respectively. Here, a denotes the vector of expansion coefficients and p denotes the vector of geometric parameters of the relative primitives. The level-set function $\Phi(x,d)$ is hence defined with the R_0-function as

$$
\begin{aligned}
\Phi(x, d) &= \Phi_{TVM}(x, a) \wedge \Phi_{FDDM}(x, p) \\
&= \Phi_{TVM}(x, a) + \Phi_{FDDM}(x, p) - \sqrt{\Phi_{TVM}(x, a)^2 + \Phi_{FDDM}(x, p)^2}
\end{aligned}
\tag{5.52}
$$

where $d = \{a, p\}$. In view of Φ_{TVM} defined in Eq. (5.7), the sensitivity of Φ is expressed as

$$
\begin{cases}
\dfrac{\partial \Phi}{\partial a_j} = \varphi_j \left(1 - \dfrac{\Phi_{TVM}}{\sqrt{\Phi_{TVM}^2 + \Phi_{FDDM}^2}} \right) \\[4mm]
\dfrac{\partial \Phi}{\partial p_k} = \dfrac{\partial \Phi_{FDDM}}{\partial p_k} \left(1 - \dfrac{\Phi_{FDDM}}{\sqrt{\Phi_{TVM}^2 + \Phi_{FDDM}^2}} \right)
\end{cases}
\tag{5.53}
$$

where φ_j is the jth CS-RBF associated with a_j. For 2D problems (i.e., $x = [x, y]^T$), the gradient of Φ is as follows

$$\nabla \Phi = \begin{bmatrix} \dfrac{\partial \Phi}{\partial x} \\ \dfrac{\partial \Phi}{\partial y} \end{bmatrix}$$

$$= \begin{bmatrix} \dfrac{\partial \Phi_{TVM}}{\partial x} + \dfrac{\partial \Phi_{FDDM}}{\partial x} - \dfrac{\Phi_{TVM}(\partial \Phi_{TVM}/\partial x) + \Phi_{FDDM}(\partial \Phi_{FDDM}/\partial x)}{\sqrt{\Phi_{TVM}{}^2 + \Phi_{FDDM}{}^2}} \\ \dfrac{\partial \Phi_{TVM}}{\partial y} + \dfrac{\partial \Phi_{FDDM}}{\partial y} - \dfrac{\Phi_{TVM}(\partial \Phi_{TVM}/\partial y) + \Phi_{FDDM}(\partial \Phi_{FDDM}/\partial y)}{\sqrt{\Phi_{TVM}{}^2 + \Phi_{FDDM}{}^2}} \end{bmatrix}$$

$$(5.54)$$

- Design case 2: Freeform design domain with nondesignable solid features

With the R_0-function, the level-set function Φ could be formulated as

$$\Phi(\pmb{x}, \pmb{d}) = \Phi_{TVM}(\pmb{x}, \pmb{a}) \vee \Phi_{FDDM}^{aux}(\pmb{x}, \pmb{q}) \wedge \Phi_{FDDM}^{basic}(\pmb{x}, \pmb{p})$$

$$= \Phi_{TVM}(\pmb{x}, \pmb{a}) + \Phi_{FDDM}^{aux}(\pmb{x}, \pmb{q}) + \Phi_{FDDM}^{basic}(\pmb{x}, \pmb{p}) + \sqrt{\Phi_{TVM}(\pmb{x}, \pmb{a})^2 + \Phi_{FDDM}^{aux}(\pmb{x}, \pmb{q})^2}$$

$$- \sqrt{\left(\Phi_{TVM}(\pmb{x}, \pmb{a}) + \Phi_{FDDM}^{aux}(\pmb{x}, \pmb{q}) + \sqrt{\Phi_{TVM}(\pmb{x}, \pmb{a})^2 + \Phi_{FDDM}^{aux}(\pmb{x}, \pmb{q})^2} \right)^2 + \Phi_{FDDM}^{basic}(\pmb{x}, \pmb{p})^2}$$

$$(5.55)$$

Similarly, the sensitivities of Φ can be calculated as

$$\begin{cases} \dfrac{\partial \Phi}{\partial a_j} = \varphi_j \left(1 + \dfrac{\Phi_{TVM}}{\sqrt{\Phi_{TVM}{}^2 + \Phi_{FDDM}^{aux}{}^2}} \right) \left(1 - \dfrac{A}{\sqrt{A^2 + \Phi_{FDDM}^{basic}{}^2}} \right) \\[4mm] \dfrac{\partial \Phi}{\partial p_k} = \dfrac{\partial \Phi_{FDDM}^{basic}}{\partial p_k} \left(1 - \dfrac{\Phi_{FDDM}^{basic}}{\sqrt{A^2 + \Phi_{FDDM}^{basic}{}^2}} \right) \\[4mm] \dfrac{\partial \Phi}{\partial q_l} = \dfrac{\partial \Phi_{FDDM}^{aux}}{\partial q_l} \left(1 + \dfrac{\Phi_{FDDM}^{aux}}{\sqrt{\Phi_{TVM}{}^2 + \Phi_{FDDM}^{aux}{}^2}} \right) \left(1 - \dfrac{A}{\sqrt{A^2 + \Phi_{FDDM}^{basic}{}^2}} \right) \end{cases}$$

$$(5.56)$$

where

$$A = \Phi_{TVM} + \Phi_{FDDM}^{aux} + \sqrt{\Phi_{TVM}{}^2 + \Phi_{FDDM^2}^{aux}} \qquad (5.57)$$

For 2D problems, the gradient of Φ is calculated as

$$
\nabla \Phi =
\begin{bmatrix}
\dfrac{\partial \Phi}{\partial x} \\[2mm]
\dfrac{\partial \Phi}{\partial y}
\end{bmatrix}
$$

$$
=
\begin{bmatrix}
B_x\left(1 - \dfrac{A}{\sqrt{A^2 + {\Phi_{FDDM}^{basic}}^2}}\right) + \dfrac{\partial \Phi_{FDDM}^{basic}}{\partial x}\left(1 - \dfrac{\Phi_{FDDM}^{basic}}{\sqrt{A^2 + {\Phi_{FDDM}^{basic}}^2}}\right) \\[5mm]
B_y\left(1 - \dfrac{A}{\sqrt{A^2 + {\Phi_{FDDM}^{basic}}^2}}\right) + \dfrac{\partial \Phi_{FDDM}^{basic}}{\partial y}\left(1 - \dfrac{\Phi_{FDDM}^{basic}}{\sqrt{A^2 + {\Phi_{FDDM}^{basic}}^2}}\right)
\end{bmatrix}
$$

$$(5.58)$$

with

$$
\begin{cases}
B_x = \dfrac{\partial \Phi_{TVM}}{\partial x} + \dfrac{\partial \Phi_{FDDM}^{aux}}{\partial x} + \dfrac{\Phi_{TVM}\left(\partial \Phi_{TVM}/\partial x\right) + \Phi_{FDDM}^{aux}\left(\partial \Phi_{FDDM}^{aux}/\partial x\right)}{\sqrt{{\Phi_{TVM}}^2 + {\Phi_{FDDM}^{aux}}^2}} \\[5mm]
B_y = \dfrac{\partial \Phi_{TVM}}{\partial y} + \dfrac{\partial \Phi_{FDDM}^{aux}}{\partial y} + \dfrac{\Phi_{TVM}\left(\partial \Phi_{TVM}/\partial y\right) + \Phi_{FDDM}^{aux}\left(\partial \Phi_{FDDM}^{aux}/\partial y\right)}{\sqrt{{\Phi_{TVM}}^2 + {\Phi_{FDDM}^{aux}}^2}}
\end{cases}
\quad (5.59)
$$

The design domain Ω_{FDDM} is a limited region within which the material layout has to be topologically optimized. This conditioned design can easily be achieved by the SIMP method because design variables are directly associated with the finite element model that fits the design domain. Instead, the conventional level-set-based topology optimization is to set manually the level-set normal derivative to zero along the design domain boundary $\partial \Omega_{FDDM}$ (Wang et al., 2003) to avoid the boundary violation, that is, the avoidance of material growth outside the design domain. This, however, will result in the so-called stopping issue (Rong and Liang, 2008).

In our implementation, because the resulting Φ in Eq. (5.17) is always negative outside Ω_{FDDM}, the zero level-set movements of Φ are confined to the design domain Ω_{FDDM} exactly. Meanwhile, the sensitivity analysis discussed below indicates that any zero level-set movement has no tendency to yield the boundary violation of the design domain. To make things clear, let us consider Fig. 5.16 as an example. It belongs to design case 1 without nondesignable solid features.

The following relations then hold

$$
\begin{cases}
\Phi_{TVM}(x) = 0 \text{ and } \Phi_{FDDM}(x) = 0 & \forall x \in \partial \Omega \cap (\partial \Omega_{TVM} \cap \partial \Omega_{FDDM}) \\
\Phi_{TVM}(x) > 0 \text{ and } \Phi_{FDDM}(x) = 0 & \forall x \in \partial \Omega \cap (\partial \Omega_{FDDM} \setminus (\partial \Omega_{TVM} \cap \partial \Omega_{FDDM})) \\
\Phi_{TVM}(x) = 0 \text{ and } \Phi_{FDDM}(x) > 0 & \forall x \in \partial \Omega \cap (\Omega_{FDDM} \setminus \partial \Omega_{FDDM})
\end{cases}
$$

$$(5.60)$$

In the current case, the boundary of Ω_{FDDM} is supposed to be unchangeable with $d = a$. The substitution of Eq. (5.60) into Eq. (5.53) then yields

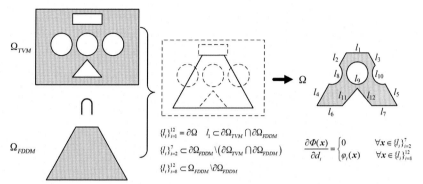

FIG. 5.16 Sketch map of the characteristics related to the different parts of the structural boundary.

$$\frac{\partial \Phi(x)}{\partial d_i} = \frac{\partial \Phi(x)}{\partial a_i} = \begin{cases} 0 & \forall x \in \partial\Omega \cap (\partial\Omega_{FDDM} \setminus (\partial\Omega_{TVM} \cap \partial\Omega_{FDDM})) \\ \varphi_i(x) & \forall x \in \partial\Omega \cap (\Omega_{FDDM} \setminus \partial\Omega_{FDDM}) \end{cases} \quad (5.61)$$

The above equation is interpreted as follows: (i) if a structural boundary point x lies on the boundary of the design domain and $\Phi_{TVM}(x)$ is positive (i.e., $x \in \{l_i\}_{i=2}^{7}$ in Fig. 5.16), the corresponding $\partial\Phi(x)/\partial d_i$ is 0; (ii) if a structural boundary point x lies in the interior of the design domain (i.e., $x \in \{l_i\}_{i=8}^{12}$ in Fig. 5.16), the calculated $\partial\Phi(x)/\partial d_i$ is the same as in the conventional method that simply adopts Φ_{TVM} in the topology optimization (Luo et al., 2008, 2009).

The sensitivity $\partial\Phi/\partial d_i$ obtained in design case 2 can be proven to have the same property by just taking A defined in Eq. (5.57) as Φ_{TVM} used in design case 1. This sensitivity property is important to essentially limit the zero level-set movements of Φ within the predefined freeform design domains. Take part of the zero level-set, l_1, in Fig. 5.16 as an example. If it has the tendency to move outside Ω_{FDDM} with Φ_{TVM} becoming positive on l_1, the movement of l_1 will automatically cease to be driven because of $\partial\Phi(x)/\partial d_i = 0$ for $x \in l_1$ according to the above analyses. As a result, the value of Φ_{TVM} on l_1 will stop growing.

Example 5.6 Numerical verification of the sensitivity property with design domain preserving.
Consider a short cantilever beam shown in Fig. 5.17 to highlight the sensitivity property introduced in this section. The rectangular design domain is partitioned into 80×50 cells as used in Rong and Liang (2008). The problem of minimizing the mean compliance J with the volume constraint $V \leq 0.768\,\mathrm{m}^3$, i.e., the volume ratio constrained to be 0.3 is investigated (Rong and Liang, 2008).

The implementation of the current topology optimization procedure is made in the following ways. A number of circular holes are prescribed within the design domain as the initial topology shown in Fig. 5.18A. The relevant level-set functions Φ_{TVM} and Φ_{FDDM} are implicit functions of the perforated domain and the rectangular design domain, respectively. Specifically, Φ_{TVM}

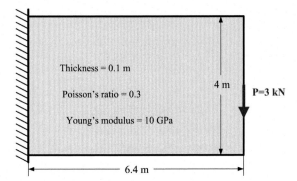

FIG. 5.17 Design domain of a cantilever beam.

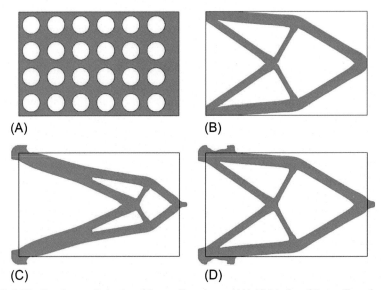

FIG. 5.18 Topology optimization of the cantilever beam: (A) initial design of the cantilever beam; (B) final design obtained by the present method ($J = 0.252189$ Nm, $V = 0.767996$ m^3); final designs obtained by the conventional method when the sensitivity $\partial \Phi(x)/\partial d_i(x \in \partial\Omega \cap \partial\Omega_{FDDM})$ is (C) not revised ($J = 0.280583$ Nm, $V = 0.767954$ m^3); or (D) directly taken as 0 ($J = 0.253583$ Nm, $V = 0.767991$ m^3).

is constructed by interpolation of CS-RBFs centered at the cell vertex with their support radius equal to 0.3 m. The level-set function Φ is defined by Eq. (5.17).

Three different final designs are plotted in Fig. 5.18B and D, and the design domain boundary is represented by black solid lines. It can be seen from Fig. 5.18B that the present method of using Φ for the description of the structural domain not only limits the zero level-set movements within the design domain accurately, but also yields the best design result with the lowest compliance value. Unfortunately, the conventional method of adopting Φ_{TVM} for the definition of the structural domain leads to results with the boundary violation of the design domain. In fact, if the sensitivity $\partial\Phi/\partial d_i$ was not artificially corrected as zero value along the design domain boundary, the conventional method would actually search for the optimum topology in an unlimited design domain and result in an unsatisfied design, as shown in Fig. 5.18C. As the remedy, Fig. 5.18D shows the design result with the artificial setting of $\partial\Phi/\partial d_i=0$ along the design domain boundary. However, this zero setting operation is hard to do manually when the shape of the design domain is irregular and changeable.

5.3.4 Hamilton-Jacobi equation for the unification of implicit and parametric formulations

Moreover, it is interesting to investigate the relationship of the sensitivities calculated with level-set and parametric formulations. In this section, attempts are made to perform shape sensitivity analysis of a structure with parametric boundary representation. The level-set method is extended and the unification of both methods is established within the framework of the fixed mesh. In view of Eq. (5.42), it is clear that the key to shape sensitivity analysis is to compute two essential terms: the inverse of the gradient module $1/\|\nabla\Phi\|$ and the sensitivity of the implicit function $\partial\Phi/\partial d_i$. Assume now that only the parametric boundary representation is available in the form of Eq. (2.5) instead of $\Phi(\boldsymbol{x},\boldsymbol{d})$. By differentiating both sides of $\Phi(\boldsymbol{x},\boldsymbol{d})=0$ with respect to d_i, we can obtain the so-called Hamilton-Jacobi equation that dominates the boundary shape evolution.

$$\frac{\partial\Phi(\boldsymbol{x},\boldsymbol{d})}{\partial d_i}+\frac{\partial\boldsymbol{x}}{\partial d_i}\cdot\nabla\Phi(\boldsymbol{x},\boldsymbol{d})=0 \qquad (5.62)$$

Based on Eq. (2.21), the aforementioned equation can further be detailed for 2D problems as

$$\frac{\partial\Phi}{\partial d_i}-\left(\frac{\partial x_1(t,\boldsymbol{d})}{\partial d_i},\frac{\partial x_2(t,\boldsymbol{d})}{\partial d_i}\right)\|\nabla\Phi\|\boldsymbol{n}=0 \qquad (5.63)$$

In the above formulation, the negative sign indicates the normal $\boldsymbol{n}$ points to the outward of the structural boundary. In view of Eq. (5.42), the required sensitivity term is then derived from Eq. (5.63).

$$\begin{aligned}\frac{\partial\Phi}{\partial d_i}\frac{1}{\|\nabla\Phi\|}&=\left(\frac{\partial x_1(t,\boldsymbol{d})}{\partial d_i},\frac{\partial x_2(t,\boldsymbol{d})}{\partial d_i}\right)\boldsymbol{n}\\[2mm]&=\frac{\dfrac{\partial x_1(t,\boldsymbol{d})}{\partial d_i}\dfrac{\partial x_2(t,\boldsymbol{d})}{\partial t}-\dfrac{\partial x_2(t,\boldsymbol{d})}{\partial d_i}\dfrac{\partial x_1(t,\boldsymbol{d})}{\partial t}}{\sqrt{\left(\dfrac{\partial x_1(t,\boldsymbol{d})}{\partial t}\right)^2+\left(\dfrac{\partial x_2(t,\boldsymbol{d})}{\partial t}\right)^2}}\end{aligned} \qquad (5.64)$$

This expression has, in fact, a sound physical meaning and can be interpreted as the projection of the location variation of a boundary point x onto the outward normal direction n, that is, the negative gradient direction whenever a perturbation of design variable d_i takes place. If $\Phi(x, d)$ concerns a distance function, this expression measures the distance variation. By substituting Eq. (5.64) into Eq. (5.42), the parametric method then produces the following sensitivity analysis formula.

$$\frac{\partial \Psi(\Phi)}{\partial d_i} = \int_{\partial\Omega} f(x) \frac{\dfrac{\partial x_1(t, d)}{\partial d_i}\dfrac{\partial x_2(t, d)}{\partial t} - \dfrac{\partial x_2(t, d)}{\partial d_i}\dfrac{\partial x_1(t, d)}{\partial t}}{\sqrt{\left(\dfrac{\partial x_1(t, d)}{\partial t}\right)^2 + \left(\dfrac{\partial x_2(t, d)}{\partial t}\right)^2}} d\Gamma$$

$$= \int_{\partial\Omega} f(x) \left(\frac{\partial x_1(t, d)}{\partial d_i}\frac{\partial x_2(t, d)}{\partial t} - \frac{\partial x_2(t, d)}{\partial d_i}\frac{\partial x_1(t, d)}{\partial t}\right) d\Gamma$$

(5.65)

Similarly, the sensitivity of the boundary cell stiffness matrix and the volume are computed by

$$\begin{cases} \dfrac{\partial k_c}{\partial d_i} = \displaystyle\int_{\partial\Omega_c} B^T DB \left(\dfrac{\partial x_1(t, d)}{\partial d_i}\dfrac{\partial x_2(t, d)}{\partial t} - \dfrac{\partial x_2(t, d)}{\partial d_i}\dfrac{\partial x_1(t, d)}{\partial t}\right) d\Gamma \\ \dfrac{\partial v_c}{\partial d_i} = \displaystyle\int_{\partial\Omega_c} \left(\dfrac{\partial x_1(t, d)}{\partial d_i}\dfrac{\partial x_2(t, d)}{\partial t} - \dfrac{\partial x_2(t, d)}{\partial d_i}\dfrac{\partial x_1(t, d)}{\partial t}\right) d\Gamma \end{cases}$$

(5.66)

In view of Eq. (5.64), the same computing scheme can be generalized to 3D problems so that

$$\frac{\partial \Phi}{\partial d_i}\frac{1}{\|\nabla\Phi\|} = \left(\frac{\partial x_1(s, t, d)}{\partial d_i}, \frac{\partial x_2(s, t, d)}{\partial d_i}, \frac{\partial x_3(s, t, d)}{\partial d_i}\right) n$$

(5.67)

Based on the above discussions, it is concluded that the parametric form of Eq. (2.5) can also be used as easily as the level-set form for shape sensitivity analysis. Although the shape sensitivity analysis formulation is developed from the level-set form of Eq. (5.42), the final computing scheme of Eq. (5.65) can be achieved in a unified way. This is a breakthrough to bridge the parametric boundary representation and the fixed mesh. In other words, the implicitization, that is, the conversion of the parametric into the level-set form, is not necessary in shape sensitivity analysis when the fixed mesh is used.

Example 5.7 Shape sensitivity analysis of an ellipse area.
In order to demonstrate the unification of both sensitivity analysis formulations, consider an ellipse whose level-set and parametric functions are as follows.

$$\Phi = 1 - \left(\frac{x^2}{a^2} + \frac{y^2}{b^2}\right)$$

(5.68)

$$\begin{cases} x = a\cos\theta \\ y = b\sin\theta \end{cases} \quad 0 \le \theta < 2\pi$$

(5.69)

where a and b denote the semilength and semiwidth of the ellipse.

Suppose a is selected as the shape design variable. According to Eq. (5.42), the implicit formulation gives rise to the following sensitivity of the ellipse area

$$\frac{\partial S}{\partial a} = \int_{\partial \Omega} \frac{\partial \Phi(x,y,a)}{\partial a} \frac{1}{\|\nabla \Phi(x,y,a)\|} d\Gamma = \int_{\partial \Omega} \frac{b^2 x^2}{a\sqrt{b^4 x^2 + a^4 y^2}} d\Gamma \qquad (5.70)$$

Based on Eq. (5.65), the parametric method gives rise to the following sensitivity of the ellipse area.

$$\frac{\partial S}{\partial a} = \int_{\partial \Omega} \frac{b\cos\theta\cos\theta + 0a\sin\theta}{\sqrt{(-a\sin\theta)^2 + (b\cos\theta)^2}} d\Gamma = \int_{\partial \Omega} \frac{b\cos^2\theta}{\sqrt{(a\sin\theta)^2 + (b\cos\theta)^2}} d\Gamma \qquad (5.71)$$

Both results are identical if Eq. (5.69) is substituted into Eq. (5.70). In case of a circle ($a=b$), Eqs. (5.42) and (5.65) produce the identical result.

$$\frac{\partial S}{\partial a} = 2\pi a \qquad (5.72)$$

Example 5.8 Shape sensitivity analysis associated with a B-spline.

Consider a quadratic uniform B-spline. Its parametric function is expressed as

$$\begin{cases} x = \frac{1}{2}\left[(x_i + x_{i+1}) + (-2x_i + 2x_{i+1})u + (x_i - 2x_{i+1} + x_{i+2})u^2\right] \\ y = \frac{1}{2}\left[(y_i + y_{i+1}) + (-2y_i + 2y_{i+1})u + (y_i - 2y_{i+1} + y_{i+2})u^2\right] \end{cases} \quad 0 \le u \le 1$$

$$(5.73)$$

where (x_i, y_i), (x_{i+1}, y_{i+1}), (x_{i+2}, y_{i+2}) denote the coordinates of three control points of the curve. (x,y) is an arbitrary point on the curve and u represents the intrinsic parameter of the curve.

With the help of the elimination theory (Sederberg et al., 1984; Walker, 1950; Yalcin et al., 2003) introduced in Section 2.4.1, the aforementioned parametric form can equivalently be converted into the implicit form

$$\Phi = b_2^2 x^2 - 2a_2 b_2 xy + a_2^2 y^2 + \left(-2a_0 b_2^2 + a_1 b_1 b_2 - a_2 b_1^2 + 2a_2 b_0 b_2\right)x +$$
$$\left(-2b_0 a_2^2 + b_1 a_1 a_2 - b_2 a_1^2 + 2b_2 a_0 a_2\right)y + a_0^2 b_2^2 + a_2^2 b_0^2 + a_1^2 b_0 b_2 + a_0 a_2 b_1^2 -$$
$$a_0 a_1 b_1 b_2 - a_1 a_2 b_0 b_1 - 2a_0 a_2 b_0 b_2$$

$$(5.74)$$

with the coefficients

$$\begin{cases} a_0 = \frac{1}{2}(x_i + x_{i+1}), a_1 = -x_i + x_{i+1}, a_2 = \frac{1}{2}(x_i - 2x_{i+1} + x_{i+2}) \\ b_0 = \frac{1}{2}(y_i + y_{i+1}), b_1 = -y_i + y_{i+1}, b_2 = \frac{1}{2}(y_i - 2y_{i+1} + y_{i+2}) \end{cases} \quad (5.75)$$

and control points

$$\begin{cases} (x_i, y_i) = (5, -30) \\ (x_{i+1}, y_{i+1}) = (18, 40) \\ (x_{i+2}, y_{i+2}) = (25, -50) \end{cases} \tag{5.76}$$

If coordinate x_i of control point i is considered as one design variable, the implicit formulation produces the computing result

$$\frac{\partial \Phi}{\partial x_i} \frac{1}{\|\nabla \Phi\|} = \frac{80xy - 2065y - 3650x - 3y^2 + 76825}{\sqrt{(18y - 480x + 16220)^2 + (480y - 12800x + 202900)}} \tag{5.77}$$

Based on Eq. (5.64), the parametric formulation gives rise to the following result

$$\begin{aligned} \frac{\partial \Phi}{\partial x_i} \frac{1}{\|\nabla \Phi\|} &= \frac{\dfrac{\partial x(u, x_i)}{\partial x_i}\dfrac{\partial y(u, y_i)}{\partial u} - \dfrac{\partial y(u, y_i)}{\partial x_i}\dfrac{\partial x(u, x_i)}{\partial u}}{\sqrt{\left(\dfrac{\partial x(u, x_i)}{\partial u}\right)^2 + \left(\dfrac{\partial y(u, y_i)}{\partial u}\right)^2}} \\ &= \frac{(1 - 2u + u^2)(70 - 160u)}{2\sqrt{(13 - 6u)^2 + (70 - 160u)^2}} \end{aligned} \tag{5.78}$$

Both results are found to be identical if Eq. (5.73) is substituted into Eq. (5.77). Fig. 5.19A shows the agreement between the parametric and level-set boundary

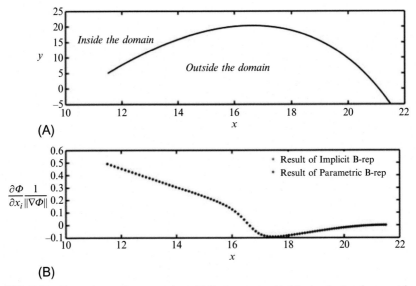

(A)

(B)

FIG. 5.19 Comparisons of curves and sensitivity terms computed by level-set and parametric methods: (A) agreement between the parametric and implicit B-reps; and (B) agreement between sensitivity terms related to Eqs. (5.77) and (5.78).

representations while Fig. 5.19B shows the agreement between sensitivity terms related to Eqs. (5.77) and (5.78), respectively.

References

Chen, J., Shapiro, V., Suresh, K., Tsukanov, I., 2007. Shape optimization with topological changes and parametric control. Int. J. Numer. Methods Eng. 71 (3), 313–346.

James, K.A., Hansen, J.S., Martins, J.R., 2009. Structural topology optimization for multiple load cases using a dynamic aggregation technique. Eng. Optim. 41 (12), 1103–1118.

Kim, N.H., Chang, Y., 2005. Eulerian shape design sensitivity analysis and optimization with a fixed grid. Comput. Methods Appl. Mech. Eng. 194 (30−33), 3291–3314.

Le, C., Norato, J., Bruns, T., Ha, C., Tortorelli, D., 2010. Stress-based topology optimization for continua. Struct. Multidiscip. Optim. 41 (4), 605–620.

Lee, E., James, K.A., Martins, J.R., 2012. Stress-constrained topology optimization with design-dependent loading. Struct. Multidiscip. Optim. 46 (5), 647–661.

Luo, Z., Wang, M.Y., Wang, S., Wei, P., 2008. A level set-based parameterization method for structural shape and topology optimization. Int. J. Numer. Methods Eng. 76 (1), 1–26.

Luo, Z., Tong, L., Kang, Z., 2009. A level set method for structural shape and topology optimization using radial basis functions. Comput. Struct. 87 (7–8), 425–434.

Luo, Y., Wang, M.Y., Kang, Z., 2013. An enhanced aggregation method for topology optimization with local stress constraints. Comput. Methods Appl. Mech. Eng. 254, 31–41.

Rong, J.H., Liang, Q.Q., 2008. A level set method for topology optimization of continuum structures with bounded design domains. Comput. Methods Appl. Mech. Eng. 197 (17–18), 1447–1465.

Sederberg, T.W., Anderson, D.C., Goldman, R.N., 1984. Implicit representation of parametric curves and surfaces. Comput. Vis. Graph. Image Process. 28 (1), 72–84.

Van Miegroet, L., 2012. Generalized Shape Optimization Using XFEM and Level Set Description. (Doctoral dissertation), Université de Liège, Liège, Belgique.

Walker, R.J., 1950. Algebraic Curves. Princeton University Press, Princeton, New Jersey.

Wang, S., Wang, M.Y., 2006. Radial basis functions and level set method for structural topology optimization. Int. J. Numer. Methods Eng. 65 (12), 2060–2090.

Wang, M.Y., Wang, X., Guo, D., 2003. A level set method for structural topology optimization. Comput. Methods Appl. Mech. Eng. 192 (1–2), 227–246.

Yalcin, H., Unel, M., Wolovich, W., 2003. Implicitization of parametric curves by matrix annihilation. Int. J. Comput. Vis. 54 (1–3), 105–115.

Zhang, W.H., Beckers, P., Fleury, C., 1995. A unified parametric design approach to structural shape optimization. Int. J. Numer. Methods Eng. 38 (13), 2283–2292.

Chapter 6

Feature-driven optimization method and applications

6.1 Unification of implicit and parametric shape optimization

Shape optimization is one of the most challenging problems in the community of structural optimization. The aim is to improve structural performance through changing the boundary shape. At an earlier stage, shape optimization was commonly carried out by virtue of the parametric boundary representation (B-rep) of a structure. Typical examples concern parameterized CAD curves/surfaces such as B-splines and NURBS in most applications where shape design variables are directly associated with control points (Braibant and Fleury, 1984). Clearly, this method has a great flexibility in CAD modeling for the geometric control of moving boundaries and also has a sound physical meaning for the definition of design variables. However, its implementation is commonly made in combination with the finite element method (FEM) where the model updating should generate a body-fitted mesh conformal to the boundary shape variation for sensitivity analysis and structure reanalysis after each iteration. Statistically, remeshing and modeling constitute about 80% of the work of the whole design task. Therefore, how to avoid remeshing and sensitivity inaccuracy constitutes the main challenges in driving the shape optimization procedure correctly. Until now, remarkable progress has been achieved (Braibant and Fleury, 1984; Kim et al., 2002; Bletzinger et al., 2010; Zhang et al., 2010; Le et al., 2011).

The fixed grid technique (García-Ruíz and Steven, 1999a,b) was introduced into shape optimization without the need for velocity field computing. To do this, a structure is enveloped by a rectangular base domain that is fully discretized with a finite element mesh. The mesh is assumed to be fixed in the sensitivity analysis and iteration process. Elements are thus classified into inner, outer, and boundary elements cut through by the domain boundary. Material properties are then averaged in terms of the solid area fraction to compute the stiffness matrices of the boundary elements. Dunning et al. (2011) proposed a weighting function in terms of both the solid area fraction and the distance from the sampling point to the boundary. However, the optimization

The Feature-driven Method for Structural Optimization. https://doi.org/10.1016/B978-0-12-821330-8.00006-7

convergence might be worsened due to the inaccuracies of local stresses and stress sensitivities. For this reason, the use of a fixed grid was limited to the minimization problem of structural compliance. García-Ruíz and Steven (1999b) improved the stress accuracy by employing a fixed grid global/local analysis with a refined local mesh around the boundary. Kim and Chang (2005) and Wang and Zhang (2013) developed a material perturbation method using the fixed finite element mesh for shape sensitivity analysis.

Compared with the parametric method, the implicit method is an alternative approach strongly adapted to the new emerging computing methods working with fixed mesh such as the extended finite element method (XFEM) (Sukumar et al., 2000; Van Miegroet and Duysinx, 2007), the meshfree methods (Luo et al., 2012), and the finite cell method (FCM) (Parvizian et al., 2007, 2012). It can easily identify whether an element is inside the solid domain according to the sign of the implicit function and can easily be used to formulate shape sensitivity analysis independently of the mesh by means of the Heaviside function and the boundary integral scheme (Wang et al., 2003; Allaire et al., 2004). As the existence of multiple material phases or a mixed solid-void phase within an element is allowed, the combination of the implicit method with the fixed mesh is straightforward for shape and even topology optimization with possible topological merging and breaking (Sethian and Wiegmann, 2000; Luo et al., 2008; Wang et al., 2015).

Actually, two forms of implicit functions, i.e., discrete and continuous level-set functions, are available for the boundary representation of a structure. In the discrete level-set method, two different sets of meshes are normally involved: a finite difference grid for propagation of the level-set function, and a finite element mesh for computing field variables. The boundary shape of a structure is grid-related with the level-set values defined at related nodes so that a large number of design variables are unavoidably brought out (Wang et al., 2003). However, the mathematical form of the level-set function lacks a concise geometric interpretation for the boundary representation and the definition of design variables. Compared with the CAD parametric curves, the implicit method cannot take advantage of the control points of CAD parametric curves, such as B-spline and NURBS in the definition of design variables. The boundary shape control is less convenient than using CAD parametric curves in practical applications. The continuous level-set function can usually be defined in the forms of the KS-function and the R-function (Shapiro, 1991; Fougerolle et al., 2005) that are constructed by means of Boolean operations of the basic geometric primitives of a structure.

It should be mentioned that both the parametric and implicit functions are, in fact, equivalent in nature but of different mathematical forms. In this sense, the current section makes it possible to easily carry out shape optimization using a parametric B-rep without mesh updating and even the boundary shape overlap is allowed for the topological change of a structure. To clarify the above presentation, three basic approaches are schematically illustrated in Fig. 6.1.

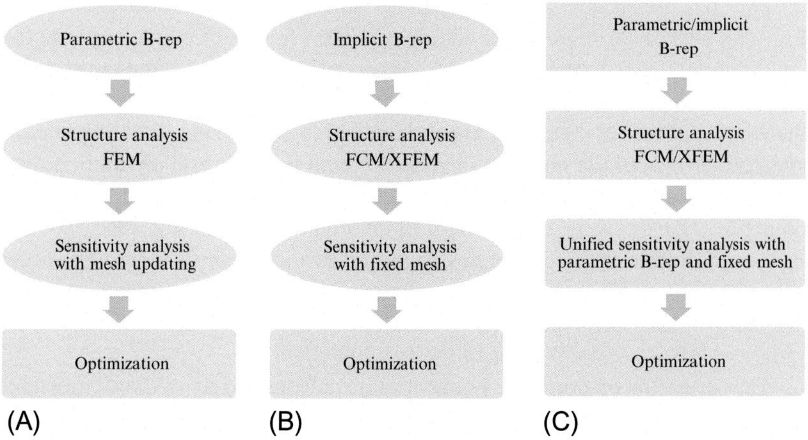

FIG. 6.1 Illustration of shape optimization approaches: (A) parametric method; (B) implicit method; and (C) unified method.

6.1.1   Implicit shape optimization with level-set functions

In this section, we introduce a shape optimization framework based on fixed mesh and level-set functions. Parametric cubic splines are expressed as the zero level-set function to describe the structural boundaries, offering great design flexibility with only the coordinates of a few interpolation points being design variables. R-functions are further utilized to combine the implicit cubic splines with other primitive level-set functions of a complex structure.

Suppose shape optimization concerns a portion of boundary $\partial\Omega$ and the physical domain Ω is located below $\partial\Omega$. A series of interpolation points $\{x_i, y_i\}$ $(i=0, 1, \ldots, n)$ is prescribed in advance. $Y = \{y_0, y_1, \ldots, y_n\}$ is considered the set of the shape design variables while x_i $\{i=0, 1, \ldots, n\}$ are considered constants and sequenced in strict monotonic order. The corresponding cubic spline interpolation function is then expressed in piecewise form as

$$y = S(x, Y) = \begin{cases} s_0(x, Y) & x_0 \leq x \leq x_1 \\ s_1(x, Y) & x_1 \leq x \leq x_2 \\ \vdots & \vdots \\ s_{n-1}(x, Y) & x_{n-1} \leq x \leq x_n \end{cases} \quad (6.1)$$

where

$$s_i(x, Y) = a_i(Y)x^3 + b_i(Y)x^2 + c_i(Y)x + d_i(Y) \quad i = 0, 1, \ldots, n-1 \quad (6.2)$$

In view of the definition in Eq. (2.12), the following level-set function can be converted from Eq. (6.1)

$$\Phi(x, y, Y) = \begin{cases} s_0(x, Y) - y & x_0 \leq x \leq x_1 \\ s_1(x, Y) - y & x_1 \leq x \leq x_2 \\ \vdots & \vdots \\ s_{n-1}(x, Y) - y & x_{n-1} \leq x \leq x_n \end{cases} \quad (6.3)$$

Fig. 6.2 illustrates the cubic spline and its implicit form defined by Eqs. (6.1)–(6.3). A cubic spline that interpolates four data points and satisfies condition $S'(x_0+0, Y) = S'(x_n - 0, Y) = 0$ is considered as the test example. The plots show that the cubic spline changes along with the design variables Y, and the zero level-sets of $\Phi(x, y, Y)$ are all identical to $S(x, Y)$. Fig. 6.3 illustrates an alternative form of the cubic spline interpolation function and its implicit form $\Phi(x, y, X)$ suitable to the situation where the y coordinates of interpolation points $\{x_i, y_i\}$ are sequenced in strict monotonic order. In this case, $X = \{x_0, x_1, \cdots, x_n\}$ defines the vector of design variables and y_i $\{i = 0, 1, \cdots, n\}$ are considered constants. The plots of the explicit and implicit form of the corresponding cubic spline interpolating four data points and satisfying condition $S'(X, y_0+0) = S'(X, y_n - 0) = 0$ confirm the validity.

The structure of complex boundaries to be optimized is then embedded within a fictitious regular domain that is discretized with fixed regular cells. The adapted integration scheme is further used to deal with cutting boundary cells. Here, the B-spline finite cell method is employed for shape optimization to achieve a high-order continuity and stress accuracy along cell boundaries with fewer degrees of freedom. The Web method is adapted to the implementation of homogeneous Dirichlet boundary conditions because the level-set

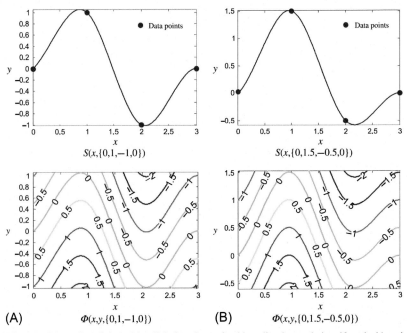

FIG. 6.2 Plots of explicit and implicit functions of cubic spline interpolating $\{0, y_0\}$, $\{1, y_1\}$, $\{2, y_2\}$ and $\{3, y_3\}$: (A) the set of design variables $\{y_0, y_1, y_2, y_3\} = \{0,1,-1,0\}$; and (B) the set of design variables changes to $\{y_0, y_1, y_2, y_3\} = \{0,1.5,-0.5,0\}$.

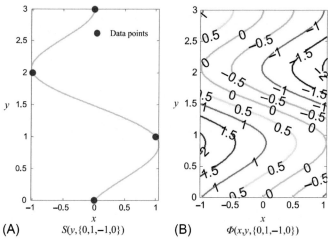

FIG. 6.3 Cubic spline interpolating $\{x_0, 0\}$, $\{x_1, 1\}$, $\{x_2, 2\}$ and $\{x_3, 3\}$: (A) explicit function $S(y, X)$ with $X = \{x_0, x_1, x_2, x_3\} = \{0, 1, -1, 0\}$; and (B) the contour map of its implicit form.

function representing the structural boundary can easily be revised as weight functions to B-spline basis functions in the interpolation of the displacement field.

The proposed method is now applied to solve stress constrained shape optimization problems. As the level-set function is adopted to describe the structural boundaries, shape optimization can allow topological changes. In this book, the globally convergent method of moving asymptotes (GCMMA) (Svanberg, 1995) within the optimization platform Boss-Quattro (Radovcic and Remouchamps, 2002) is used as the optimizer.

Example 6.1 Implicit shape optimization of a torque arm.

The torque arm shown in Fig. 6.4 was originally studied by Bennett and Botkin (1985) by using a parametric B-rep in combination with the finite element method. In their work, the finite elements were updated to keep conformity with the boundary variation. It is now considered a benchmark by many researchers (Zhang et al., 1995; Kim and Chang, 2005; Van Miegroet, 2012). The left circle of the torque arm is fixed. A horizontal load of 2789 N and a vertical load of 5066 N are simultaneously applied at the center of the right circle. The Young's modulus and Poisson's ratio are 207.4 GPa and 0.3, respectively. The thickness of the torque arm is 3 mm. The optimization problem related to Eq. (5.19) is solved. The von Mises stress is contrained to be smaller than 800 MPa.

Both inner and outer boundaries are supposed to be designable. As shown in Fig. 6.4, the inner designable boundary is controlled by four shape design variables. The outer boundary is modeled by cubic curves and has six design variables associated with y-coordinates of the independent control points (in red) to

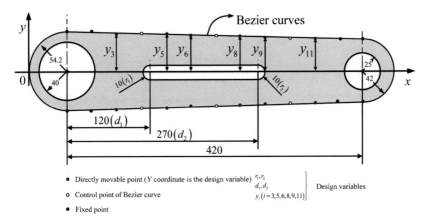

FIG. 6.4 The dimensions, boundary conditions, and design variables of the torque arm.

keep the structural symmetry. Notice that the empty points representing the dependent control points are determined as the middle points of two neighboring independent control points to ensure the first-order continuity of the moving boundary while the solid points in blue denote the fixed control points. The level-set function is constructed for the torque arm by means of R-functions as well as cubic splines $\Phi(x,y,Y)$, as in Example 5.1.

The finite cell model of the torque arm is constructed, as shown in Fig. 6.5. Notice that cells marked in red denote the boundary cells that will further be partitioned into subcells by the quadtree approach. The fixation of the left circle shown in Fig. 6.5 corresponds to the homogeneous Dirichlet boundary condition. A simple approximation might fill the left circle with a solid material largely stiffer than the material of the torque arm and then fix the control points related to the B-spline cells inside the left circle. Unfortunately, the numerical

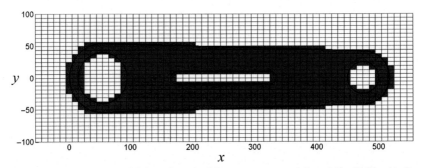

FIG. 6.5 Zero level-set and finite cell mesh of the torque arm consisting of 61×32 bi-cubic B-spline cells.

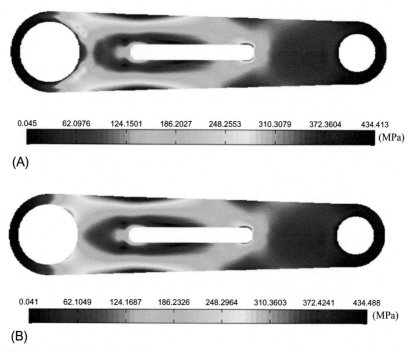

FIG. 6.6 Contour plots of von Mises stresses obtained by the B-spline finite cell method with two different modeling methods of the Dirichlet boundary condition: (A) a stiff material filling the left circle; and (B) implementation of the Web method.

results given in Fig. 6.6A indicate that the accuracy of the von Mises stress around the fixed circle is greatly deteriorated. With the help of the Web method discussed in Section 4.2, the left circle can exactly be fixed owing to the effect of the weight function ω. Fig. 6.6B shows that a more precise result of the von Mises stress can be obtained around the left circle.

During the optimization process, free variations of the design variables might change the strucutral topology with the left circle cut by the interior slot, as shown in Fig. 6.7.

To avoid one such intersection, the following additional constraint is introduced for topology preserving.

$$80 - r_1 + x_c \geq \delta \qquad (6.4)$$

where x_c denotes the variation of the x-coordinate of the circle center $\boldsymbol{Q}$ marked in Fig. 5.3A and $\delta = 3$ denotes the tolerance value used to control the minimum distance between the right end point $\boldsymbol{P}$ of the left circle and the left end point of the interior slot.

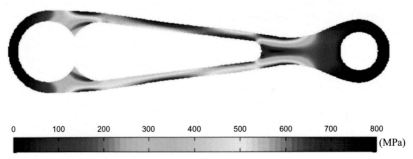

0   100   200   300   400   500   600   700   800
(MPa)

FIG. 6.7 Contour plot of von Mises stress with the interior slot and the left circle merging together.

Generally speaking, the boundary varaition can effectively be controlled by means of the level-set function during the optimization process. If the variation of the zero-value contour $\Phi(x_\tau)=0$ is limited within a tolerance band defined by lower and upper distance bounds $\underline{d}$ and $\bar{d}$, the constraints are then expressed as

$$\Phi\left(x_\tau - \bar{d}\frac{\nabla\Phi(x_\tau)}{\|\nabla\Phi(x_\tau)\|}\right) \leq 0, \quad \Phi\left(x_\tau + \underline{d}\frac{\nabla\Phi(x_\tau)}{\|\nabla\Phi(x_\tau)\|}\right) \geq 0 \qquad (6.5)$$

Numerically, a set of discrete points of x_τ can be selected and substituted into the above conditions for approximation. According to the work of (Alexandrov and Santosa, 2005), the logarithmic barrier method can be used as the interior penalty to satisfy these constraints. With this idea in mind, the minimum gap, i.e., the distance between two varying boundaries, could also be controlled. Suppose $\Phi_1(x)=0$ and $\Phi_2(x)=0$ are implicit functions representing two separated boundaries of a structure or two objects. $d_{\min}$ and $d_{\max}$ are the allowable minimum and maximum distances between both. $x_i^{(1)}$ and $x_i^{(2)}$ are two such points located on both boundaries that measure the minimum distance between $\Phi_1(x)=0$ and $\Phi_2(x)=0$. According to Taylor's expansion, the distance can be approximated by

$$d = \left\|x_i^{(1)} - x_i^{(2)}\right\| = -\frac{\Phi_1\left(x_i^{(2)}\right)}{\left\|\nabla\Phi_1\left(x_i^{(2)}\right)\right\|} = -\frac{\Phi_2\left(x_i^{(1)}\right)}{\left\|\nabla\Phi_2\left(x_i^{(1)}\right)\right\|} \qquad (6.6)$$

Consequently, one can introduce the following constraint to control d

$$d_{\min} \leq d \leq d_{\max} \qquad (6.7)$$

The corresponding level-set function and the contour plot of the von Mises stresses after optimization are shown in Figs. 6.8A–B, respectively. The optimized result using the parametric method in combination with FEM is also shown in Fig. 6.8C. It is indicated that a similar result can be obtained with the proposed method. Fig. 6.9 gives the iteration histories of the maximum stress and the objective function (volume) with a reduction of more than 50%.

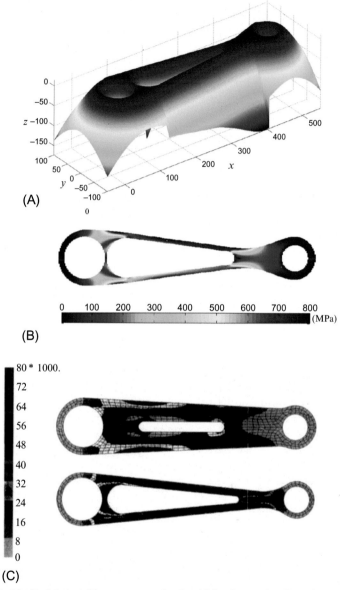

FIG. 6.8 The final design of the torque arm using the additional constraint of Eq. (6.4): (A) level-set function constructed by R-functions; (B) contour plot of von Mises stress; and (C) optimized result using parametric method in combination with FEM (Zhang et al., 1995).

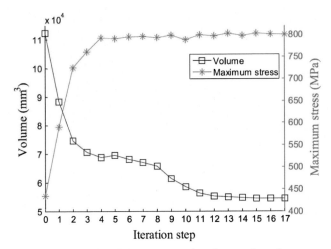

FIG. 6.9 Convergence histories of the torque arm's volume and maximum stress with the implicit method.

Example 6.2 Implicit shape optimization of a bracket.

The bracket model is shown in Fig. 6.10. It was initially studied by Bennett and Botkin (1985) for shape optimization. Material properties, the upper bound of stress constraints, and the kind of optimization problem are the same as in the previous example. The outer boundary excluding circular arcs is modeled by pieces of cubic splines in terms of level-set functions to change the shape. The inner triangular hole is optimized together. Suppose the two circles at the bottom are fixed. A horizontal force of 15,000 N is applied at the center of the upper circle.

Fig. 6.11A shows the finite cell model of the bracket. The von Mises stress contour plot obtained by the B-spline finite cell method is shown in Fig. 6.11B.

Similarly, an additional constraint is introduced to avoid the overlap between the inner triangular contour and the two fixed bottom holes such that

$$\sqrt{(50 - d_1)^2 + d_2^2} - (r_1 + 10) \geq \delta \tag{6.8}$$

The final design is shown in Fig. 6.12 with $\delta = 5$. The bracket topology is even changed with the successful removal of the bottom branch due to its weak stress intensity. Fig. 6.13 shows that a gain of more than 60% reduction of the

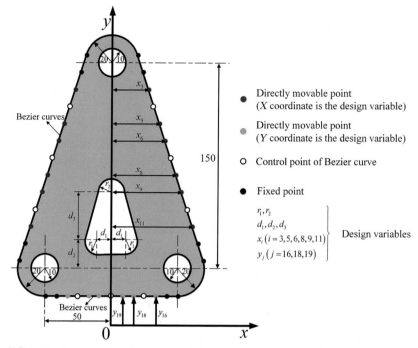

FIG. 6.10   The dimensions, boundary conditions, and design variables of the bracket.

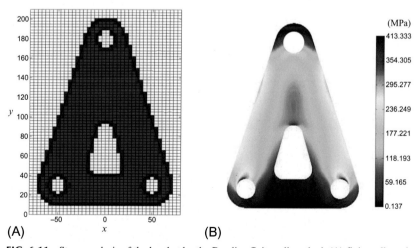

FIG. 6.11   Stress analysis of the bracket by the B-spline finite cell method: (A) finite cell mesh consisting of 40×70 bi-cubic B-spline cells; and (B) contour plot of von Mises stress.

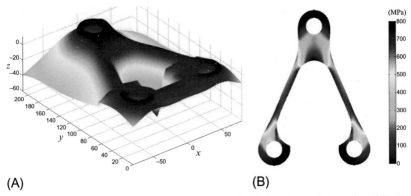

(A)   (B)

FIG. 6.12 The optimum design of the bracket obtained with extra constraint equation (6.8): (A) corresponding level-set representation; and (B) contour plot of von Mises stress.

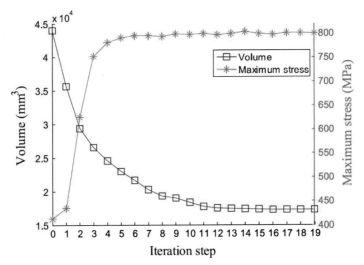

Iteration step

FIG. 6.13 Convergence histories of the bracket's volume and maximum stress with the implicit method.

structural volume is achieved after a smooth iteration history of the objective function (volume) without stress violation. In fact, the structural volume would be further reduced if the radii (20 mm in Fig. 6.10) related to the three arcs of the outer boundary are relaxed as design variables.

6.1.2 Unified shape optimization with parametric functions and fixed mesh

Parametric and implicit methods are traditionally thought to be two irrelevant approaches in structural shape optimization. This section will unify both implicit and parametric methods to break through the separation situation. It is shown that the combination of the parametric method with the fixed mesh greatly favors the design flexibility and structural reanalysis after each shape modification. To do this, the parametric boundary representation is first discretized into polygons and then analyzed using the popular point-in-polygon test, also called the inclusion test in Section 2.4.2, to identify the inner, outer, and boundary elements of a structure. The finite cell method is implemented to perform structural analysis under the fixed mesh where high-order B-spline functions are adopted as shape functions to interpolate the overall displacement field. Then, by means of the Hamilton-Jacobi equation, the formulation of Eq. (5.65) derived from adapting the implicit method to the parametric method is adopted for analytic shape sensitivity analysis formulation.

In this section, the shape optimizations of the torque arm and bracket structures are further studied without the implicitization of the parametric descriptions. The motivation is to show that the combination of the parametric B-rep with the finite cell method is indeed an efficient approach to achieve structural shape optimization. As the mesh is fixed and not needed to update no matter how the structure changes, this new approach will reduce the gap between the shape optimization methodology and the practical design needs.

Example 6.3 Unified shape optimization of a torque arm.
The shape optimization problem shown in Fig. 6.4 is further studied with the parametric B-rep with the fixed mesh of the finite cell method. The combination greatly favors both geometric modeling and structural analysis. The element types shown in Fig. 6.14 are determined with the ray-crossing method.

Fig. 6.15A and B show the von Mises stresses related to the initial and final shape, respectively. Fig. 6.16 indicates the locations of the control points after optimization. The dependent control points are depicted by empty points. Fig. 6.17 clearly indicates that the iteration has a good convergence with a weight reduction of more than 50%. In comparison with the results in Fig. 6.8B and C, the different methods are capable of achieving nearly the same results.

Example 6.4 Unified shape optimization of a bracket.

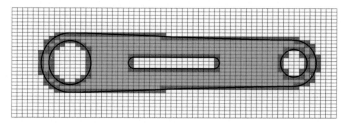

FIG. 6.14   Finite cell model of the torque arm and cell classifications.

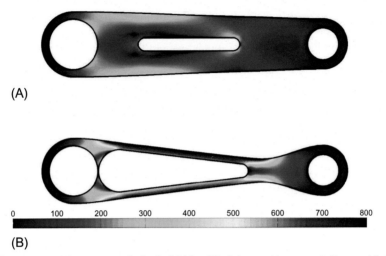

(A)

(B)

FIG. 6.15 Von Mises stress results for the initial and final shapes with parametric B-rep and finite cell analysis: (A) initial shape; and (B) final shape.

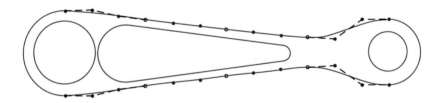

- Directly movable point (*Y* coordinate is the design variable)
- Control point of Bezier curve
- Fixed point

FIG. 6.16 Locations of control points after optimization.

The shape optimization of the bracket in Fig. 6.10 is also studied. The finite cell model obtained with the ray-crossing method is shown in Fig. 6.18. Boundary overlapping is allowed to make changes in the structural topology during shape optimization. This implies that complete material removal is possible from the boundary crossing domain.

The final positions of the control points are shown in Fig. 6.19 after optimization. The design evolution of the structure is given in Fig. 6.20. It is seen that a weight reduction of nearly 60% is achieved for the structure.

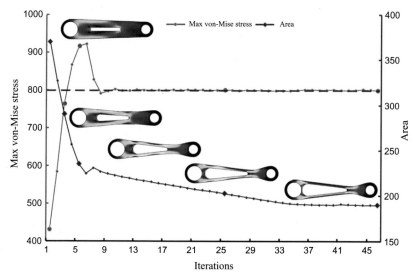

FIG. 6.17 Convergence histories of the torque arm's volume and maximum stress with the unified method.

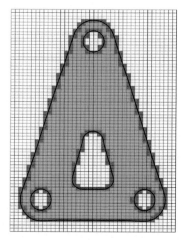

FIG. 6.18 Finite cell model of the bracket obtained with the ray-crossing method.

6.2 Shape optimization of the Dirichlet and free boundaries

In most existing works (Braibant and Fleury, 1984; Bennett and Botkin, 1985; Laurent-Gengoux and Mekhilef, 1993; Zhang et al., 1995; Haslinger and Mäkinen, 2003), the standard shape optimization only concerns free boundaries of a structure while the Dirichlet boundaries are assumed to be unchangeable.

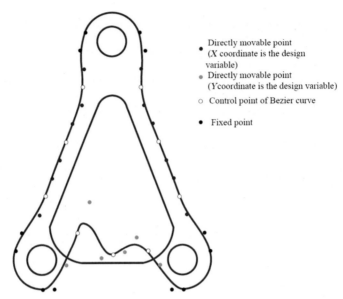

FIG. 6.19 Locations of control points after optimization.

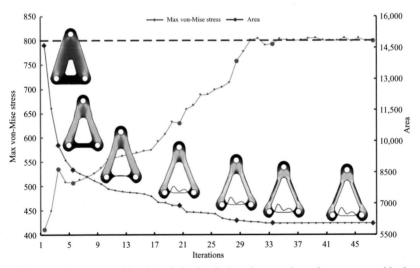

FIG. 6.20 Convergence histories of the bracket's volume and maximum stress with the unified method.

In fact, the design of structural supports constitutes a long-term motivating research topic involved in many structural engineering fields such as building construction, workpiece machining fixtures, welding, and fastener connections or rivet joints of aircraft structures. Because any adjustment of support positions may considerably influence the structural performance, the optimization of boundary/support conditions plays a crucial role and should be carefully designed.

Historically, the optimization of elastic support positions mainly concerned 1D beam-like structures to improve eigenvalues and dynamic performances (Rozvany, 1974; Olhoff, 1988). Later, the idea of topology optimization using the SIMP model was extended into the support position design and even the simultaneous optimization of both supports and 2D/3D structures. In (Buhl, 2002; Zhu and Zhang, 2006), elastic supports are modeled as a certain number of discrete springs to realize the compliance minimization and natural frequency maximization, respectively. Multicomponent layout design was dealt with in (Zhu and Zhang, 2010), where supports are considered elastic components and allowed to be relocated continuously along the specific boundaries of a structure. An interesting approach was developed in (Xia et al., 2014), where the shape and topology optimization of both the structure and the Dirichlet boundaries was carried out by means of the level-set method for compliance minimization. The mesh updating was, however, obliged to maintain a body-fitted finite element mesh along with the design modification of the Dirichlet boundaries. Especially, the discrete level-set function form used to control the geometric shape of the Dirichlet boundaries unavoidably brings out a large number of design variables that are mesh-dependent.

In this section, an efficient shape optimization method of the Dirichlet boundaries is proposed by combining the weighted B-spline finite cell method with the level-set function. It is shown that the underlying problems related to the large number and mesh-dependence of design variables, the discretization of the DBC, and mesh updating can be completely avoided. According to the Web method, the DBC is satisfied by penalizing the displacement field with the weighting function defined by the level-set function. This means that the Dirichlet boundary shapes are described by implicit representations with continuous design variables independent of the mesh discretization. Meanwhile, the finite cell method is implemented to use a fixed grid over the whole optimization process, no matter how the structural shape changes. As is seen, this method will provide great flexibility in generating the discrete mesh independently of the physical domain of a structure so that discrete nodes are not necessarily located on the structural boundaries.

Consider the shape optimization formulations in Eqs. (5.19) and (5.21). The sensitivity of the structural mean compliance $J(u, \Phi)$ is calculated as

$$\frac{\partial J}{\partial d_i} = U^T K \frac{\partial U}{\partial d_i} + \frac{1}{2} U^T \frac{\partial K}{\partial d_i} U \tag{6.9}$$

Based on the stress-strain relation

$$\sigma = D\varepsilon = D\overline{B}U \tag{6.10}$$

Stress sensitivity is then calculated as

$$\frac{\partial \sigma}{\partial d_i} = D\frac{\partial \overline{B}}{\partial d_i}U + D\overline{B}\frac{\partial U}{\partial d_i} \tag{6.11}$$

in which sensitivities of $\overline{B}$ and U are calculated as follows. According to Eq. (4.37), we have

$$\frac{\partial \overline{B}}{\partial d_i} = \begin{bmatrix} \dfrac{\partial \overline{B}_1}{\partial d_i} & \cdots & \dfrac{\partial \overline{B}_r}{\partial d_i} & \cdots & \dfrac{\partial \overline{B}_{m\times n}}{\partial d_i} \end{bmatrix}$$

$$= \begin{bmatrix} \dfrac{\partial \omega}{\partial d_i}\dfrac{\partial M_r}{\partial x} + M_r\dfrac{\partial^2 \omega}{\partial x \partial d_i} & 0 \\[3mm] \cdots & \dfrac{\partial \omega}{\partial d_i}\dfrac{\partial M_r}{\partial y} + M_r\dfrac{\partial^2 \omega}{\partial y \partial d_i} & \cdots \\[3mm] \dfrac{\partial \omega}{\partial d_i}\dfrac{\partial M_r}{\partial y} + M_r\dfrac{\partial^2 \omega}{\partial y \partial d_i} & \dfrac{\partial \omega}{\partial d_i}\dfrac{\partial M_r}{\partial x} + M_r\dfrac{\partial^2 \omega}{\partial x \partial d_i} \end{bmatrix} \tag{6.12}$$

According to the equilibrium equation $KU=F$, we have

$$\frac{\partial U}{\partial d_i} = K^{-1}\left(\frac{\partial F}{\partial d_i} - \frac{\partial K}{\partial d_i}U\right) \tag{6.13}$$

Clearly, the core is to calculate the sensitivities of K and F.

- The sensitivity of global stiffness matrix K

$$\frac{\partial K}{\partial d_i} = \frac{\partial}{\partial d_i}\int_D \overline{B}^T(\omega(x,d))D\overline{B}(\omega(x,d))H(\Phi(x,d))d\Omega$$

$$= \int_D \left(\frac{\partial \overline{B}^T}{\partial d_i}D\overline{B} + \overline{B}^T D\frac{\partial \overline{B}}{\partial d_i}\right)H(\Phi)d\Omega + \int_D \overline{B}^T D\overline{B}\frac{\partial H(\Phi)}{\partial d_i}d\Omega \tag{6.14}$$

$$= 2\int_D \frac{\partial \overline{B}^T}{\partial d_i}D\overline{B}\cdot H(\Phi)d\Omega + \int_{\partial\Omega} \overline{B}^T D\overline{B}\cdot \frac{\partial \Phi}{\partial d_i}\frac{1}{\|\nabla\Phi\|}d\Gamma$$

In the above expression, the first term exists only when the design variable d_i is involved in the definition of the Dirichlet boundaries. It vanishes whenever free boundaries are only concerned with d_i.

- The sensitivity of the global load vector F

$$\frac{\partial F}{\partial d_i} = \frac{\partial}{\partial d_i}\left(\int_D \overline{M}^T(\omega(x,d))fH(\Phi(x,d))d\Omega + \int_{\Gamma_N} \overline{M}^T(\omega(x,d))td\Gamma\right)$$

$$= \int_D \frac{\partial \overline{M}^T}{\partial d_i}fH(\Phi)d\Omega + \int_D \overline{M}^T f\frac{\partial H(\Phi)}{\partial d_i}d\Omega + \int_{\Gamma_N} \frac{\partial \overline{M}^T}{\partial d_i}td\Gamma \tag{6.15}$$

$$= \int_D \frac{\partial \omega}{\partial d_i}M^T fH(\Phi)d\Omega + \int_{\partial\Omega} \overline{M}^T f\frac{\partial \Phi}{\partial d_i}\frac{1}{\|\nabla\Phi\|}d\Gamma + \int_{\Gamma_N} \frac{\partial \omega}{\partial d_i}M^T td\Gamma$$

in which the Neumann boundary Γ_N is assumed to be unchangeable in shape optimization. As the level-set function $\Phi(x, d)$ and weighting function $\omega(x, d)$ are both analytical, it is straightforward to compute their partial derivatives with respect to design variables d. While level-set functions are relatively complex, the derivatives could be calculated with the aid of automatic differentiation techniques (Rall, 1981).

Several representative examples are given below to demonstrate the capability of the developed shape optimization method. Discussions will be made about the shape optimization of the Dirichlet boundary and the simultaneous shape optimization of the free and Dirichlet boundaries.

6.2.1   Shape optimization of the Dirichlet boundary

Example 6.5 Shape optimization of the Dirichlet boundary for a curved beam. As shown in Fig. 6.21A, a curved beam is clamped at the top end and uniformly loaded by a vertical force of 6 kN on the right end. Suppose the Young's modulus and Poisson's ratio are 200 GPa and 0.29, respectively. According to Boresi et al. (1985) and Kumar et al. (2008), the maximum compression and traction stresses occur at point A and B with $\sigma_\theta^A = -4.4687$ MPa and $\sigma_\theta^B = 2.9212$ MPa, respectively.

Based on Boolean operations illustrated in Fig. 6.21B, the level-set representation of the curved beam is constructed by means of the R-function shown in Fig. 6.21C. The weighting function ω representing the fixed boundary can be constructed using Eq. (4.44). Related expressions correspond to

$$\begin{cases} f = x/100 \\ g = 1 - \left(\dfrac{y-400}{100}\right)^2 - \left(\dfrac{x}{100}\right)^2 \\ h = -f^2 \\ \omega = -\left(g + h - \sqrt{g^2 + h^2}\right) \end{cases} \qquad (6.16)$$

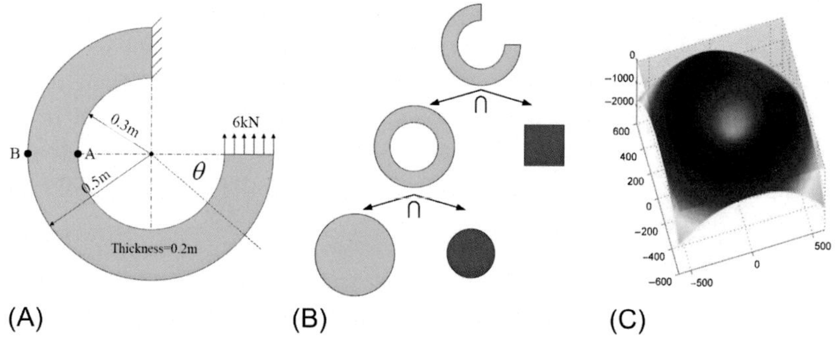

(A)   (B)   (C)

FIG. 6.21 Boolean operations defining the level-set representation of a curved beam: (A) the curved beam; (B) Boolean operations; and (C) level-set function constructed by the R-function.

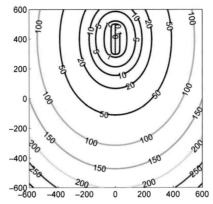

FIG. 6.22 Contour plots of weighting function defining the Dirichlet boundaries of the curved beam.

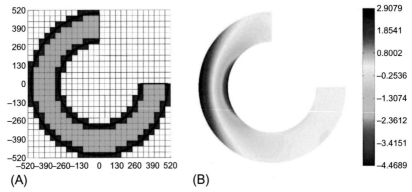

(A)   (B)

FIG. 6.23 FCM mesh and σ_y of the curved beam: (A) zero level-set and the FCM mesh of the curved beam; and (B) distribution of σ_y.

Fig. 6.22 shows that the contour plots of the weighting function ω vanish on the top end of the curved beam. Fig. 6.23A depicts the curved beam embedded into a 520 mm × 520 mm square, which is discretized into 26 × 26 finite cells. Notice that cubic B-spline basis functions are used here. Component σ_y is obtained and shown in Fig. 6.23B. The computed values of the stresses at points A and B are $\sigma_y^A = -4.4689$ MPa and $\sigma_y^B = 2.9079$ MPa with relative errors of only 0.00448% and 0.455% compared with exact solutions.

Suppose now that only two short segments with a length of 2 mm along the top end of the beam are fixed and their center positions will be optimized using Eq. (5.21). Here, d_1 and d_2 illustrated in Fig. 6.24A refer to two design variables. The volume constraint is discarded. The upper bound of the stress constraint is set to be the initial value. Figs. 6.24B and C indicate the initial and final

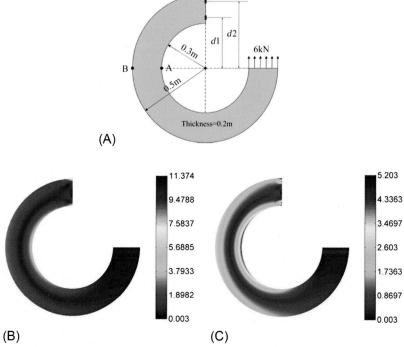

(A)

(B)   (C)

FIG. 6.24 Two-segment fixation optimization of the curved beam: (A) design variables of fixation locations; (B) initial von Mises stress; and (C) final von Mises stress.

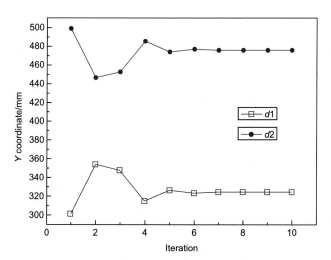

FIG. 6.25 Convergence history of fixation location design variables.

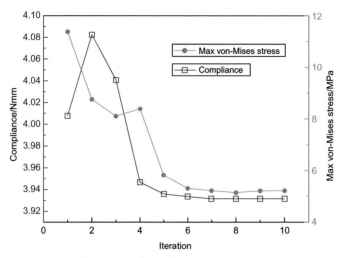

FIG. 6.26 Convergence histories of the compliance and maximum von Mises stress.

distribution of von Mises stresses. Fig. 6.25 shows that both d_1 and d_2 converge stably. Fig. 6.26 gives the convergence histories of the structural compliance and the maximum von Mises stress, from which we can see that the compliance is reduced only about 1.90% while the maximum von Mises stress is admissible and is reduced up to 54.3%.

6.2.2 Simultaneous shape optimization of free and Dirichlet boundaries

Example 6.6 Simultaneous shape optimization of the Dirichlet and free boundaries of a torque arm.

This benchmark (Fig. 6.27) was originally from a free boundary shape optimization problem and was also studied in Section 6.1. The DBC is imposed to fix the left hole. Here, the simultaneous shape optimization is applied to the inner and outer contours as well as the fixed hole shape. To apply the force at the center of the right hole, the so-called "stiff strip" (Düster et al., 2008) is imposed inside the right hole. To this end, the linear-elastic constitutive tensor $\boldsymbol{D}$ is multiplied by a scalar much larger than 10, that is, assuming the material is one order of magnitude stiffer than in the physical domain, as indicated in Fig. 6.29.

The fixed hole shape related to the Dirichlet boundary condition is now modeled as a superellipse so that the weighting function reads

$$\omega(x, y) = \left|\frac{x - 5.42}{a}\right|^{n} + \left|\frac{y}{b}\right|^{n} - 1 \tag{6.17}$$

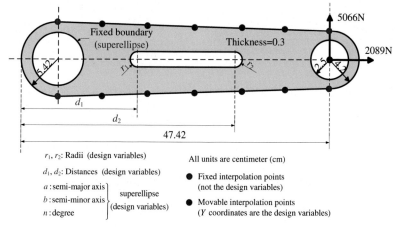

FIG. 6.27 Description of the torque arm model.

in which parameters a, b, n constitute three shape design variables whose initial values are set to be 3, 3, and 2, respectively. Meanwhile, cubic splines are used to interpolate the outer contour.

To avoid the intersection of the inner slot and the superellipse, an additional constraint is considered as

$$d_1 - 5.42 - a - r_1 \geq \delta \qquad (6.18)$$

with $\delta = 1$ in this example.

The final design is obtained after 21 iterations. The initial and final level-set functions of the torque arm are compared in Fig. 6.28. Fig. 6.29 illustrates the initial and final FCM meshes of the torque arm, respectively. Notice that the finite cells in red denote the boundary cells that are refined into subcells by quadtree refinement.

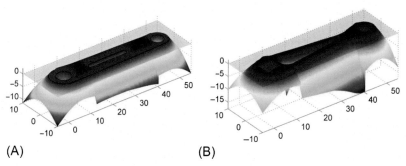

(A)　　　　　　　　　　　　　(B)

FIG. 6.28 The level-set function of the torque arm: (A) initial shape; and (B) final shape.

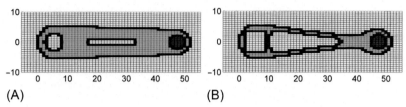

FIG. 6.29 FCM mesh of the torque arm: (A) initial shape; and (B) final shape.

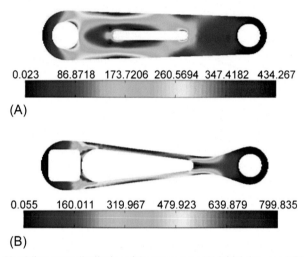

FIG. 6.30 Von Mises stress distribution of the torque arm: (A) initial shape; and (B) final shape.

Fig. 6.30 gives the von Mises stress distributions of the initial shape and final design while Fig. 6.31 illustrates the convergence history of the volume and the maximum von Mises stress of the torque arm. We can see that the maximum von Mises stress is always admissible, although it increases from the initial value of 434.267 MPa to 799.835 MPa. Instead, the structural volume is reduced by more than 50%. The fixed hole changes its circle shape into a superellipse with optimized values of $a=3.5$, $b=3.5$, and $n=7.257$.

To reveal the influence of the initial configuration of the torque arm upon the shape optimization solution, consider an alternative design case without the inner slot. Fig. 6.32 shows the von Mises stress distributions related to the initial and final shape of the torque arm. Correspondingly, Fig. 6.33 gives the convergence histories of the volume and the maximum von Mises stress. Clearly, the present solution is less effective. We can resort to the level-set based topology optimization method for more effective structures.

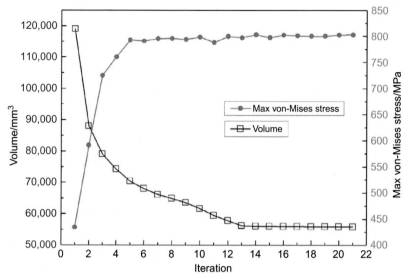

FIG. 6.31   Convergence histories of the volume and maximum von Mises stress of the torque arm.

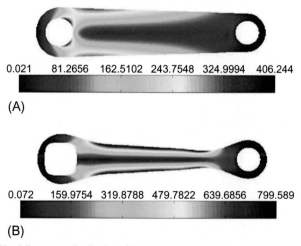

FIG. 6.32   Von Mises stress distribution of the torque arm without the inner slot: (A) initial shape; and (B) final shape.

Example 6.7 Simultaneous shape optimization of the Dirichlet and free boundaries of a bracket.

The structure is depicted in Fig. 6.34. It was originally studied to fulfill the shape optimization of the inner and outer free boundaries, excluding the three circular holes of the structure. Here, the problem is extended to perform shape optimization, including two fixed holes at the bottom.

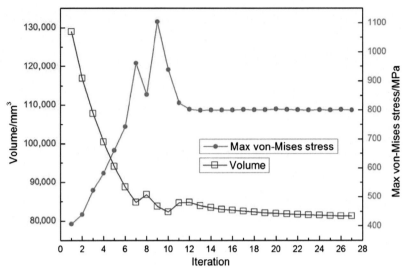

FIG. 6.33 Convergence histories of the volume and maximum von Mises stress of the torque arm without the inner slot.

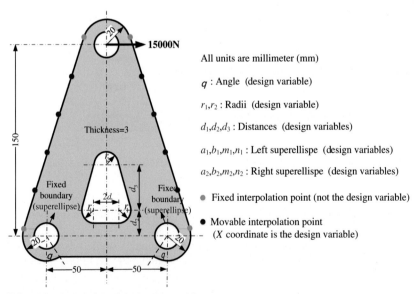

All units are millimeter (mm)

q : Angle (design variable)

r_1, r_2 : Radii (design variable)

d_1, d_2, d_3 : Distances (design variables)

a_1, b_1, m_1, n_1 : Left superellispe (design variables)

a_2, b_2, m_2, n_2 : Right superellispe (design variables)

● Fixed interpolation point (not the design variable)

● Movable interpolation point
  (X coordinate is the design variable)

FIG. 6.34 Description of the bracket model.

Suppose material properties and stress constraints are the same as in the torque arm studied before. Superellipses are still used to define weighting functions related to the Dirichlet boundary conditions of both holes.

$$\begin{cases} \omega_1(x, y) = \left|\dfrac{x+50}{a_1}\right|^{m_1} + \left|\dfrac{y-30}{b_1}\right|^{n_1} - 1 \\ \omega_2(x, y) = \left|\dfrac{x-50}{a_2}\right|^{m_2} + \left|\dfrac{y-30}{b_2}\right|^{n_2} - 1 \end{cases} \tag{6.19}$$

According to Section 3.3, the combined R-function reads

$$\omega(x, y) = \omega_1 + \omega_2 - \sqrt{\omega_1^2 + \omega_2^2} \tag{6.20}$$

The contour values are shown in Fig. 6.35. As expected, the weighting function $\omega(x, y)$ equals zero only along the two fixed holes. Geometrically, due to the impossible intersection between two superellipses, $\omega(x, y)$ is also differentiable. In order to avoid the overlap between the inner slot and the two fixed holes, the following additional constraints are introduced

$$\begin{cases} \sqrt{(50-d_1)^2 + d_2^2} - \sqrt{a_1^2 + b_1^2} - r_1 \geq \delta_1 \\ \sqrt{(50-d_1)^2 + d_2^2} - \sqrt{a_2^2 + b_2^2} - r_1 \geq \delta_2 \end{cases} \tag{6.21}$$

with $\delta_1 = \delta_2 = 4$ in this test case.

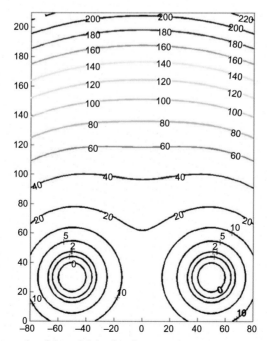

FIG. 6.35 Contour plot of the weighting function.

Comparisons are made in Figs. 6.36–6.38 to show the level-set function, FCM mesh, and von Mises stress distributions of the initial and final designs. The convergence histories of the volume and the maximum von Mises stress are shown in Fig. 6.39. It concludes that fixed holes are greatly changed in shape with $a_1=9.98$, $b_1=9.99$, $m_1=7.07$, $n_1=6.47$, $a_2=9.96$, $b_2=9.97$, $m_2=6.84$, and $n_2=6.51$. The increased maximum von Mises stress is strictly controlled by the upper bound while the structural volume is reduced about 60%.

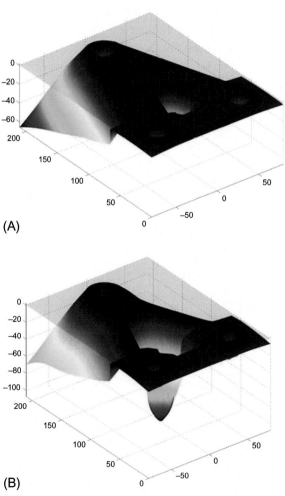

FIG. 6.36 The level-set function of the bracket model: (A) initial shape; and (B) final shape.

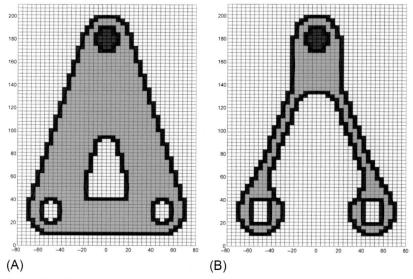

FIG. 6.37   FCM mesh of the bracket model: (A) initial shape; and (B) final shape.

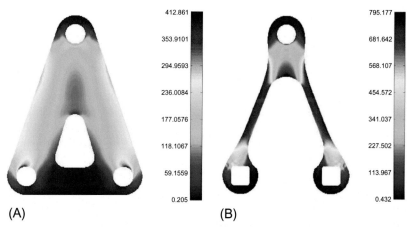

FIG. 6.38   Von Mises stress distribution in the bracket: (A) initial shape; and (B) final shape.

6.3   Topology optimization of the regular design domain

Since the work of Bendsøe and Kikuchi (1988) with the homogenization method, topology optimization has become an advanced approach for structural design and has received great attention. Various topology optimization methods have been developed in the past three decades. Recent reviews (Sigmund and Maute, 2013; Deaton and Grandhi, 2014; Zhu et al., 2016) highlight the

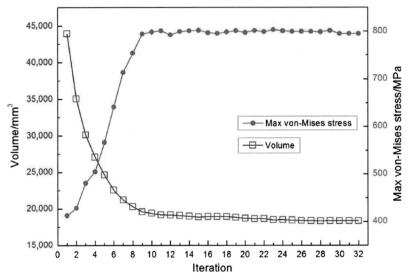

FIG. 6.39 Convergence histories of the volume and maximum von Mises stress of the bracket.

developments and applications of topology optimization methods. Generally, three kinds of methods exist.

- *Density-based method*

The density-based method has been recognized as the most popular method. In general, it is an element- or node-based method that relates the material distribution determining the structural topology to element density variables through interpolation models such as the SIMP model (Bendsøe and Sigmund, 1999), RAMP model (Stolpe and Svanberg, 2001), etc. The evolutionary structural optimization (ESO) method (Xie and Steven, 1997; Tanskanen, 2002) is also a kind of density-based method. This approach removes ineffective elements based on an intuitive criterion. Actually, the density-based method has been applied to varieties of problems including multidisciplinary design problems, owing to its conceptual simplicity and easy implementation. For this kind of method, notice that the jagged structural boundaries depend on the numerical resolution and can be improved with mesh refinement. Besides, a density/sensitivity filter is commonly used to prevent numerical instabilities such as checkerboard and mesh-dependence, which, however, introduces a gray transition from the solid to the void around the structural boundaries. However, the implicitly imposed length scale control related to the filter radius may become invalid for complex physical problems such as hinges in compliant mechanisms and thin elements in heat conductions. The intermediate densities can be minimized or entirely eliminated by the Heaviside projection method

(Guest et al., 2004) and the minimum length scale is ensured with the robust topology optimization formulation (Wang et al., 2011).

- *Level-set method*

The level-set method originated from the moving boundary tracking problem (Osher and Sethian, 1988) and was introduced into the structural optimization field for free shape and topology changes (Wang et al., 2003; Allaire et al., 2004). This method belongs to a kind of boundary-based method and works directly with the boundary of a structure for topology optimization. Actually, two forms of level-set methods are available for the topology representation of a structure. The traditional level-set method is based on the discrete level-set function that is mesh-related with the use of level-set values at discrete nodes (Wang et al., 2003; Allaire et al., 2004). This method is hindered in practical applications because of numerical complexities such as the reinitialization, velocity extension, the limitation of the Courant-Friedrichs-Lewy condition, and the huge number of design variables. Continuous LSF is usually constructed to overcome these difficulties. Wang and Wang (2006) and Luo et al. (2008) introduced the compactly supported radial basis function (CS-RBFs) to replace design variables defined by level-set values at mesh nodes with CS-RBFs' coefficients. The implicit functions of the closed form, including the KS function and the R-function, are also applied by means of Boolean operations of basic geometric primitives. In this sense, the number of design variables, the sensitivity analysis scheme, and the gray region control scheme are different between the density-based approach and the continuous level-set method. Moreover, the topological derivative method (Burger et al., 2004; Norato et al., 2007) is usually incorporated into the level-set method so that new holes can be automatically nucleated in the design domain.

- *Integrated design method of multicomponent structures*

The integrated design method of multicomponent structures is dedicated to dealing with complex structure systems. Initially, simple features were considered fixed objects incorporated within a frame structure. The level-set function was introduced by Qian and Ananthasuresh (2004) to describe the components as rigid objects. In the works of Zhang et al. (2011), the generalization of the simultaneous optimization of components and the frame structure was made possible in terms of mixed design variables, i.e., component locations/shapes and SIMP-based material densities of elements. The computing efficiency is further ameliorated with the superelement modeling of movable components, the overlapping of which is avoided by means of the so-called finite-circle method (Zhu et al., 2010). Moreover, Gao et al. (2015) treated the multicomponent structure as a double-layer system and introduced multipoint constraints to connect components with the frame structure. Alternatively, Xia et al. (2013) and Zhang et al. (2012) adopted the level-set description for the deformable solid and void components while the frame structure was described with the

SIMP-based material densities. Similarly, Kang and Wang (2013) introduced featured holes with the level-set description in combination with the frame structure described with nonlocal Shepard interpolation-based material densities. Zhou and Wang (2013) combined the traditional level-set method for the frame structure with the LSF for the features and constructed a constructive solid geometry-based level sets model. Chen et al. (2007) introduced the level-set function of B-splines instead of CS-RBF for the topology optimization of the frame structure. Mei et al. (2008) proposed a topology optimization framework that subtracts simple featured holes as geometric primitives from the design domain in consideration of the machining requirement. Liu and Ma (2015) introduced new polyline-arc features with the feature-fitting algorithm by regulating the noisy velocity fields. The features after shape optimization were subtracted from the design domain using a milling method. Recently, Norato et al. (2015) adopted bars of fixed width and semicircular ends whose signed distance function can analytically be constructed. Bars can be removed during the optimization process by penalizing the out-of-plane thickness, as in the SIMP method.

6.3.1 RBF-based topology optimization

Example 6.8 Stress-constrained topology optimization of an L-shaped beam. The structure is shown in Fig. 6.40. This test is widely considered a benchmark for stress-constrained topology optimization (Le et al., 2010; Wang and Li, 2013). Dimensionless parameters are used for material properties, load, and geometry data with the Young's modulus and Poisson's ratio being 1 and 0.3, respectively.

The initial topology and level-set function of the L-shaped beam are defined in Example 5.2. The finite cell model of the initial design is shown in Fig. 6.41A. Likewise, cells outside the design domain are naturally regarded as fictitious ones. Correspondingly, Fig. 6.41B shows the initial von Mises stress contour plot obtained by the B-spline finite cell method. It should be

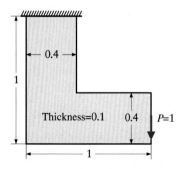

FIG. 6.40 Design domain and loading conditions of the L-shaped beam.

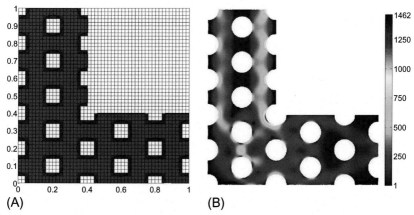

FIG. 6.41 Analysis of the initial model by the B-spline finite cell method: (A) finite cell mesh consisting 50×50 bi-cubic B-spline cells; and (B) contour plots of initial von Mises stresses.

noted that no stress concentration occurs around the application point of the external load $P = 1$ because the point-wise load is equivalently discretized over five control points in the analysis model defined by the B-spline finite cell method.

For the purpose of comparison, the problems of minimizing strain energy with and without stress constraint are investigated. The upper bound of the volume is set to be $\overline{V} = 0.02$ in both cases while the stress limit is set to be $\overline{\sigma} = 800$.

Fig. 6.42 shows the design results without a stress constraint. In detail, Fig. 6.42A gives the final topology obtained after 127 iterations. A material corner of $270°$ is generated with a high stress concentration, as shown in Fig. 6.42B. Therefore, this solution would only be reasonable for the stiffness maximization but should be avoided in strength design by the engineering intuition. Fig. 6.42C illustrates the convergence histories of the strain energy and volume. At the first few iterations, the strain energy increases sharply and then decreases smoothly because the volume constraint is violated at the beginning.

The results of the stress-constrained optimization problem are shown in Figs. 6.43 and 6.44. We can see the reentrant corner angle is no longer $270°$ in Fig. 6.43A and B. The region around the corner has a refined modification that effectively reduces the stress level. Fig. 6.44 shows the contour plots of von Mises stress at some intermediate iterations. For example, the design at iteration 50 is mostly like the final design of the previous optimization problem shown in Fig. 6.42B. Because both optimization problems have the common objective function, the angle of its reentrant corner being also $270°$ might be required for strain energy reduction. The topologies related to the designs at iterations 100, 150, and 200 have no significant changes because the strain energy already attains a low level and is not much influenced by the slow reduction of the stress concentration. Therefore, the final design given in Fig. 6.43A demonstrates sufficient strength and high stiffness.

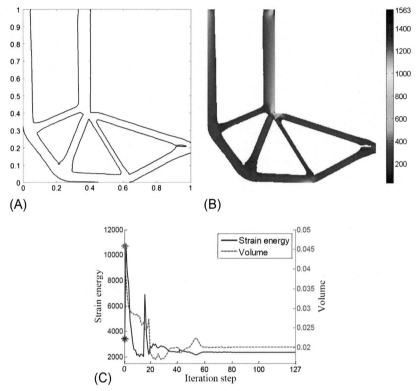

FIG. 6.42 Minimum strain energy design of the L-shaped beam without stress constraint: (A) zero level-set contour of the final design; (B) contour plot of von Mises stresses; and (C) iteration histories of the strain energy and volume.

6.3.2 Feature-driven topology optimization

Both the material densities of elements and the discrete level-set values of nodes are lower-level design variables. To some extent, they belong to a kind of free-form topology optimization where the design intent of preserving engineering features is not reflected. For numerical implementation, these approaches are also hindered by the huge number of design variables. Moreover, as the mathematical expression is not unique for an implicit function, how to select a proper form of the level-set function to represent a feature still remains a problem. As illustrated later, this problem leads to the disagreement between the material distribution model and the geometric model of the structure, which eventually results in gray regions in the optimized configuration.

In this section, we introduce a new feature-driven topology optimization (FDTO) method. A structure is considered as the union of a certain number of specific engineering features, also called components, such as mechanical parts, devices, electronic components, modular substructures, and shaped holes.

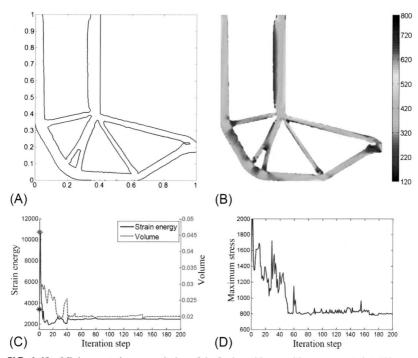

FIG. 6.43 Minimum strain energy design of the L-shaped beam with stress constraint: (A) zero level-set contour of the final design; (B) contour plot of von Mises stresses; (C) convergence histories for the strain energy and volume; and (D) convergence histories for the maximum von Mises stress.

They are designable and needed practically in view of their functionality, aesthetics, assemblage, machining facility, or other considerations. A structure consisting of three features is illustrated in Fig. 6.45. Topology optimization is driven by changing locations and/or shapes of features via Boolean operations so that the location, orientation, size, and shape parameters of features are usually chosen as design variables. Therefore, the features embedded within a structure can be partly or wholly covered by other features or reduced to a negligible size. Useless features can even vanish when they move outside the design domain. The definition of design variables is, in fact, the main difference from the standard density-based topology optimization method that is related to the FE model. Fig. 6.46 illustrates the feature-driven topology evolutions of a structure.

Boolean operation-based KS functions are hierarchically constructed to define the level-set functions that describe the engineering features and the whole structure. The narrow-band domain integral scheme in Section 5.3.1 is adopted for sensitivity analysis using a fixed regular mesh. The narrow-band domain is directly determined by a modified Heaviside function and the level-set function describing the structure. It is shown that the nonequidistant

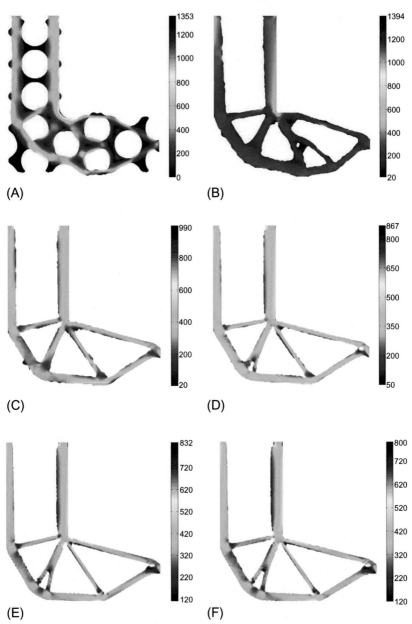

FIG. 6.44 Intermediate and final designs of the stress-constrained optimization problem: (A) iteration 5; (B) iteration 20; (C) iteration 50; (D) iteration 100; (E) iteration 150; and (F) iteration 200.

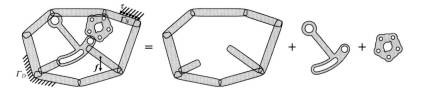

A structure   Feature 1 of superelliptical bars   Feature 2   Feature 3

FIG. 6.45   Illustration of a structure consisting of multiple engineering features.

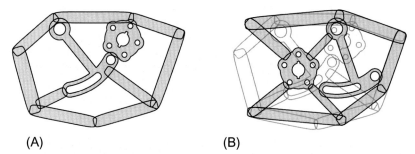

(A)   (B)

FIG. 6.46   Schematic of feature-driven topology optimization: (A) initial structure configuration; and (B) optimized structure configuration.

distribution of the narrow-band domain related to the transition interval of the modified Heaviside function critically influences the attainment of a clear pattern of the optimized configuration. More importantly, the signed distance function or its first-order approximation is found to be essential to this attainment.

Example 6.9 Topology optimization of a traditional cantilever beam.

Fig. 6.47 shows a cantilever beam loaded by a concentrated force F at the center point of the right side of the design domain. The dimension of the design domain is 80×40 and it is discretized with 80×40 quadrilateral elements. The Young's modulus of solid material is $E_0 = 1$ and Poisson's ratio is 0.3. Parameters $\Delta = 0.2$ in the smoothed Heaviside function of Eq. (5.36) and $\alpha = 1 \times 10^{-5}$

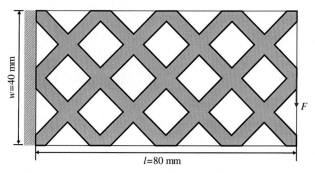

FIG. 6.47   Cantilever beam loaded at the center of the right side.

are tested. Suppose that the initial structure is composed of 16 superellipses and the KS function is used to aggregate all the level-set functions of superellipses into a single one for the whole structure. The initial length and width of each superellipse are 15 and 2, respectively. The superellipse is described with the level-set function in Eq. (2.46) and its first-order approximation of signed distance function in Eq. (2.42).

Figs. 6.48 and 6.49 show the optimization processes of the cantilever beam with the two different level-set functions. It is clear that large amounts of

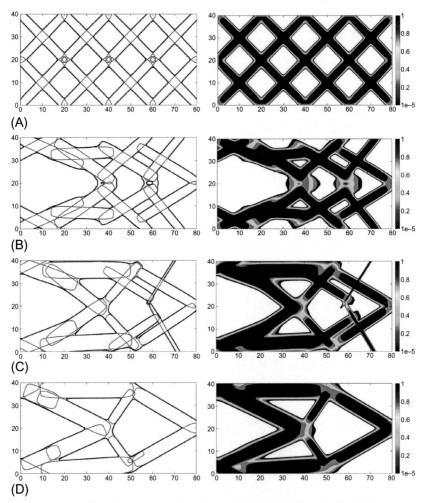

FIG. 6.48 Evolutions of structural configuration involving the level-set function of the superellipse in Eq. (2.46): (left) features *(red lines)* and the whole structure defined by the KS function *(black line)*; (right) material distribution: (A) iteration 1; (B) iteration 10; (C) iteration 29; and (D) optimized configuration.

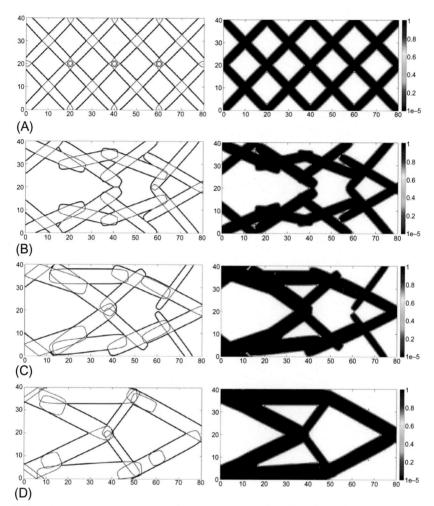

FIG. 6.49 Evolutions of structural configuration involving the first-order approximation of the superellipse in Eq. (2.42): (left) features *(red lines)* and the whole structure defined by the KS function *(black line)*; (right) material distribution: (A) iteration 1; (B) iteration 10; (C) iteration 29; and (D) optimized configuration.

intermediate materials exist near the joint portions due to the nonequidistant property of the level-set function in Eq. (2.46). When the approximated signed distance function related to Eq. (2.42) is used to describe the boundary of superellipses, intermediate materials are eliminated successfully and a clearer structural boundary is obtained. Fig. 6.50 shows that iteration curves achieve the convergence.

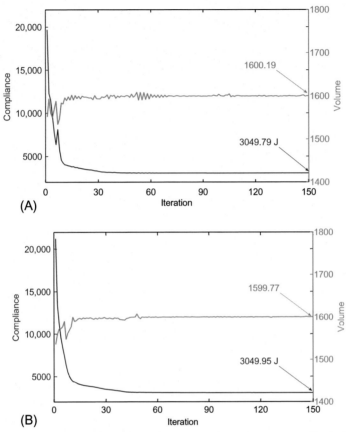

FIG. 6.50 Convergence curves of the structural compliance and volume: (A) with the level-set function of the superellipse in Eq. (2.46); and (B) with the first-order approximation of the superellipse in Eq. (2.42).

Example 6.10 Topology optimization of a cantilever beam including two specific engineering features.

A cantilever beam with two specific engineering features is considered. Suppose the Young's modulus of solid material is $E_0 = 1$ and Poisson's ratio is 0.3. Parameters $\Delta = 0.1$ and $\alpha = 1 \times 10^{-5}$ are used in Eq. (5.36). As shown in Fig. 6.51, the design domain ($l = 120$ and $w = 40$) is fixed along the left side and a concentrated force $F = 10$ is loaded on the right bottom corner. The design domain is further discretized with a regular mesh of 120×40 quadrilateral bilinear elements. This implies that 4800 pseudodensity variables would be available if the discretization of the pseudodensity field adopts one unknown for each element, as in the SIMP-based topology optimization method (Xia et al., 2013). Assume both engineering features are movable with six design variables and

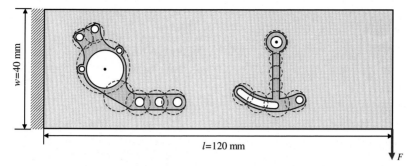

FIG. 6.51 A cantilever beam including two specific engineering features.

initially located at the reference positions (80, 30) and (20, 20), respectively. The finite circle method (Zhu et al., 2010) is adopted to avoid the geometric overlap of both features. The two engineering features are approximated with seven and eight finite circles, respectively. Two different initial configurations are tested to illustrate their influences.

For the first initial configuration, the length and width of each superellipse are $a = 18$ and $b = 2$. Suppose that 24 superellipses are distributed in a crossing way over the design domain and each of them is defined by the level-set function in Eq. (2.46) and the first-order approximation of the signed distance function in Eq. (2.42), respectively. Each superellipse has five design variables related to the position, orientation, semilength, and semiwidth. In total, 126 design variables exist. Fig. 6.52 shows the evolution of the structural configuration constructed by the KS function for which the level-set function of the superellipse in Eq. (2.46) is used. Fig. 6.53 shows the evolution of the structural configuration where the first-order approximation in Eq. (2.42) is used. The boundary of each superellipse is marked with red lines and the boundary of the whole structure obtained with the KS function is marked with a black line. Besides, the material distribution of the structure is depicted with colorful contours. The convergence histories of structural compliance and volume are given in Fig. 6.54A and B for both cases. During the optimization process, the superellipses and engineering features seek the optimal shapes, locations, and orientations. The optimized compliances are 9223.04 and 9172.14 with a decrease over 80% when the level-set function and first-order approximation of the signed distance function are used. Clearly, with the first-order approximation of the signed distance function, gray materials shown in Fig. 6.52 are well eliminated in Fig. 6.53 and the material usage satisfies the prescribed volume fraction.

For the second initial configuration, the length and width of each superellipse are $a = 12$ and $b = 2$, respectively. Suppose 32 superellipses are orthogonally placed in the design domain. In total, 166 design variables are available. Fig. 6.55 shows the evolution of the structural configuration related to the level-

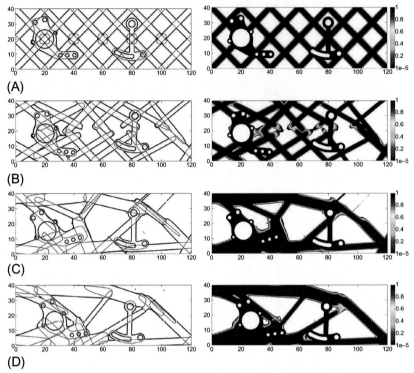

(A)

(B)

(C)

(D)

FIG. 6.52 Evolutions of structural configuration involving the level-set function of the superellipse in Eq. (2.46): (left) features *(red lines)* and the whole structure defined by the KS function *(black line)*; (right) material distribution: (A) iteration 1; (B) iteration 5; (C) iteration 25; and (D) optimized configuration.

set function of the superellipse in Eq. (2.46) while Fig. 6.56 shows the evolution of the structural configuration related to the first-order approximation of the superellipse in Eq. (2.42). Fig. 6.57 shows the convergence histories of structural compliance and volume in both cases. Likewise, with the first-order approximation of the signed distance function, the distribution of gray materials shown in Fig. 6.55 is completely eliminated in Fig. 6.56. It should be mentioned that only hundreds of design variables are available in the example to provide a great benefit for solving large-scale engineering design problems.

Moreover, the design domain is discretized with 120×40 and 240×80 elements for the first initial configuration to show the mesh-dependence problem. The optimized topologies and their convergence curves are shown in Figs. 6.58 and 6.59, respectively. Clear configurations are obtained and the final designs are slightly different in both cases. Table 6.1 indicates that only a difference of 0.68% exists for the structural compliances between the optimized topology shown in Fig. 6.58A and B when the same discretization of 240×80 elements is applied.

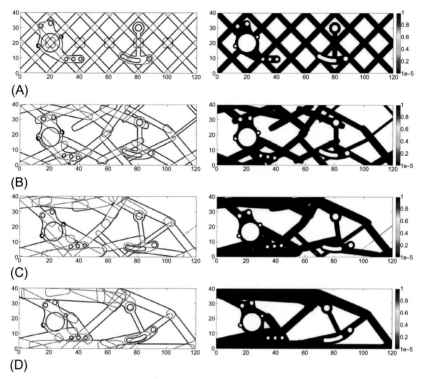

(A)

(B)

(C)

(D)

FIG. 6.53 Evolutions of structural configuration involving the first-order approximation of the superellipse in Eq. (2.42): (left) features *(red lines)* and the whole structure defined by the KS function *(black line)*; (right) material distribution: (A) iteration 1; (B) iteration 10; (C) iteration 18; and (D) optimized configuration.

6.3.3   CBS-based topology optimization

Example 6.11 Topology optimization of a short beam.

A short beam is shown in Fig. 6.60. The beam is completely fixed along the left side and is loaded by a vertical force at the bottom right corner. Suppose that the design domain is discretized into a 60×30 finite cell mesh and a volume fraction of 40% is used as the upper bound of the volume constraint in the minimization of structural compliance. A dimensionless study is made with the Young's modulus and Poisson's ratio being 1 and 0.3, respectively.

The quadratic CBS presented in Section 2.3.3 is used as a designable freeform hole to realize the topological changes of the design domain. In order to illustrate the influence of the number of design variables upon the optimized topology, three cases depicted in Table 6.2 are considered. Initial hole layouts are correspondingly shown in Fig. 6.61 together with the identical FCM mesh.

Fig. 6.62 shows the optimized topologies in all three cases and they agree well. Fig. 6.63 illustrates the convergence curves. The final values of structural

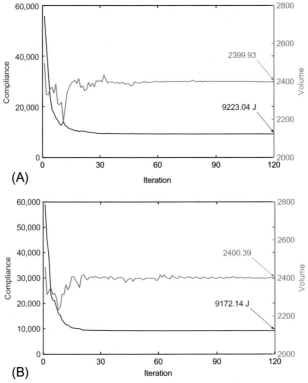

FIG. 6.54 Convergence curves of the structural compliance and volume for the first initial configuration: (A) with the level-set function of the superellipse in Eq. (2.46); and (B) with the first-order approximation of the superellipse in Eq. (2.42).

compliance correspond to 9042.34, 8772.43, and 8740.11. This means that the compliance of the optimized structure can further be reduced with increasing the number of design variables. To have a clear idea about the topological changes, some intermediate iterations are given in Table 6.3 to show the optimization process. It is observed that the overlapping of CBS and the movements of CBS outside the design domain are allowable. This problem also demonstrates that topology optimization could be carried out by means of fewer design variables of the CBS than the conventional density method.

Example 6.12 Topology optimization of a short beam involving voids or solid inclusions.

This example has the same design domain, material properties, and mesh discretization as in the previous example. Three cases shown in Fig. 6.64 are considered. Case 1 refers to a solid design domain as usual. Case 2 refers to a solid design domain involving two prescribed voids. Case 3 refers to a solid design domain involving two prescribed solid inclusions that have the same material properties as the short beam. In all three cases, suppose that two pointwise forces are applied at the center and right corner of the bottom edge.

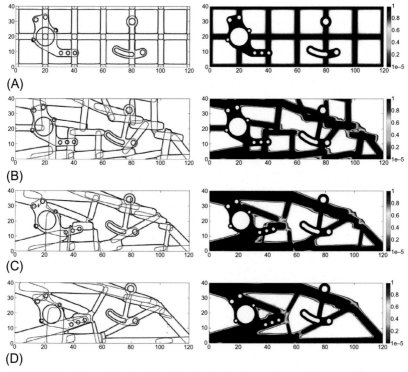

(A)

(B)

(C)

(D)

FIG. 6.55 Evolutions of structural configuration involving the level-set function of the superellipse in Eq. (2.46): (left) features *(red lines)* and the whole structure defined by the KS function *(black line)*; (right) material distribution: (A) iteration 1; (B) iteration 16; (C) iteration 50; and (D) optimized configuration.

As shown in Table 6.4, the same number of design variables related to CBSs are used in the three cases. The initial layouts of the CBS are correspondingly shown in Fig. 6.65 together with the identical finite cell mesh.

Fig. 6.66 shows that the optimized topologies are different depending upon whether voids and solid inclusions are present. Fig. 6.67 shows the corresponding convergence histories of the structural compliance and the volume. The final values of the structural compliance are 7163.33, 8267.14, and 7227.12 in the three cases. In detail, case 1 produces the structural topology with the minimal compliance because the design domain is completely free in design. Case 2 produces the structural topology with the maximal compliance because the prescribed voids obstruct the layout of the load path in the design domain. Case 3 is a compromising solution because the prescribed solid inclusions can act as the load-bearing part along the load path. Table 6.5 shows the evolutions of structural topologies in the three cases. Materials are gradually removed from the design domain through shape variations of CBS and their Boolean operations.

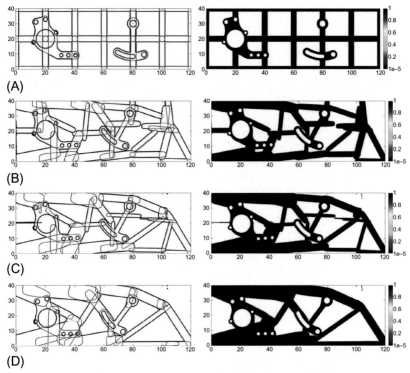

FIG. 6.56 Evolutions of structural configuration involving the first-order approximation of the superellipse in Eq. (2.42): (left) features *(red lines)* and the whole structure defined by the KS function *(black line)*; (right) material distribution: (A) iteration 1; (B) iteration 16; (C) iteration 25; and (D) optimized configuration.

6.4 Topology optimization of the freeform design domain

Topology optimization confined to freeform design domains is a challenging problem that reflects the real needs of practical applications. Compared with the so-called ground structure of regular design domain, a real structure has, in fact, a design domain limited with boundaries of free forms that are more complicated than conventionally studied rectangles, L-shapes, and others. Meanwhile, as the standard SIMP method basically resorts to the fixed finite element mesh, it is hard to ensure the stress accuracy for design domains of changeable boundaries without mesh updating. The material perturbation technique (Xia et al., 2012; Wang and Zhang, 2013) could be used to simplify the sensitivity analysis scheme.

In fact, the conventional level-set based topology optimization resorted to the discrete form in terms of nodal values of a finite element model (Wang

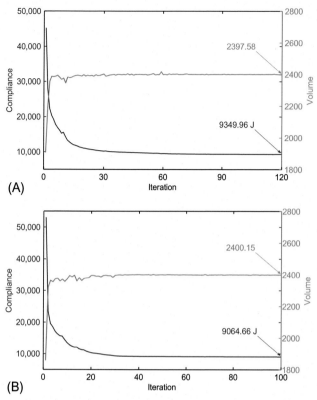

FIG. 6.57 Convergence curves of the structural compliance and volume for the second initial configuration: (A) with the level-set function of the superellipse in Eq. (2.46); and (B) with the first-order approximation of the superellipse in Eq. (2.42).

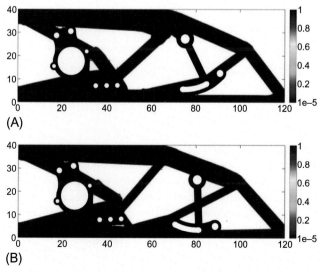

FIG. 6.58 Final designs with different mesh densities for the first initial configuration: (A) mesh 120×40; and (B) mesh 240×80.

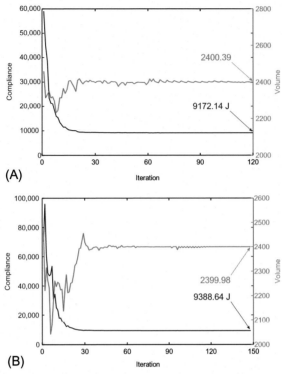

FIG. 6.59 Convergence curves with different mesh densities for the first initial configuration: (A) mesh 120 × 40; and (B) mesh 240 × 80.

TABLE 6.1 Compliance of final designs with the mesh of 240 × 80.

Final design	Fig. 6.58A	Fig. 6.58B
Compliance (J)	9325.01	9388.64

et al., 2003; Allaire et al., 2004) and could not handle freeform design domains easily because the corresponding Hamilton-Jacobi equations are solved by the finite difference scheme working only over a structured grid. Xing et al. (2010) adopted the finite element-based level-set method for irregular design domains, but it has nearly the same defect as the SIMP method when treating design domains with moving boundaries. Zhou and Wang (2013) proposed the constructive solid geometry based level-set description. Chen et al. (2007) constructed B-splines and R-function-based level-set description (Shapiro, 1991, 2007). Both works were

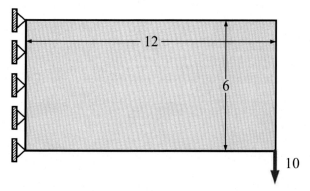

FIG. 6.60 A short beam with a vertical load at the corner.

TABLE 6.2 Three design cases for the short beam.

Items	Number of CBS	Number of design variables for each CBS	Total number of design variables
Case 1	3	24	72
Case 2	3	54	172
Case 3	17	54	918

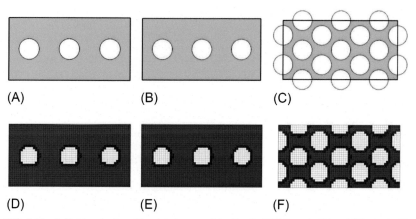

(A) (B) (C)

(D) (E) (F)

FIG. 6.61 Initial topologies of the short beam and the identical FCM mesh: (A) and (D) case 1; (B) and (E) case 2; and (C) and (F) case 3.

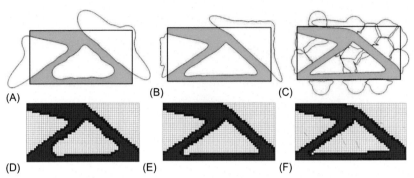

FIG. 6.62 Optimized topologies of the short beam in three cases: (A) and (D) case 1; (B) and (E) case 2; and (C) and (F) case 3.

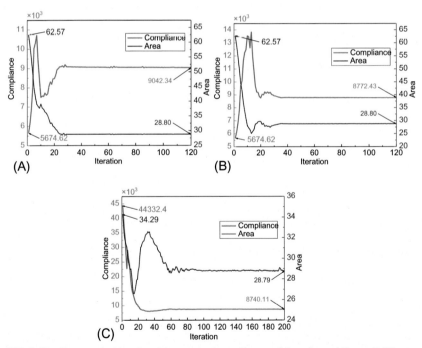

FIG. 6.63 Convergence histories of the structural compliance and the volume: (A) case 1; (B) case 2; and (C) case 3.

TABLE 6.3 Evolutions of structural topology in three cases.

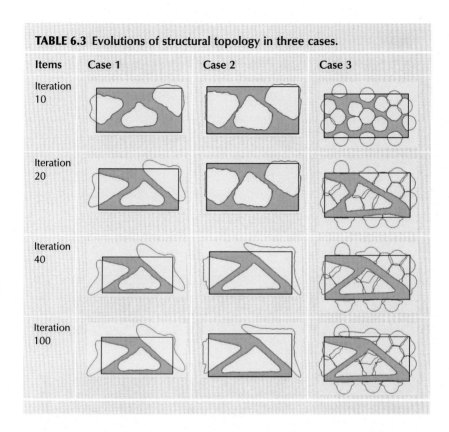

Items	Case 1	Case 2	Case 3
Iteration 10			
Iteration 20			
Iteration 40			
Iteration 100			

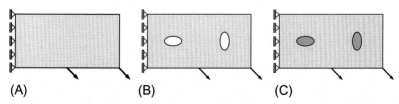

(A)   (B)   (C)

FIG. 6.64   A short beam with voids or solid inclusions: (A) case 1; (B) case 2; and (C) case 3.

only limited to the compliance-related topology optimization with freeform design domains and nondesignable solid features involved in the design domain were not considered comprehensively. James et al. (2012) and James and Martins (2012) introduced the isoparametric concept into the stress-related topology optimization for those design domains whose shape and configuration are easily constructed by mapping operations from intact rectangular regions.

TABLE 6.4 Three design cases.

Items	Number of CBS	Number of design variables for each CBS	Total number of design variables	Inclusions
Case 1	16	24	384	None
Case 2	16	24	384	2 voids
Case 3	16	24	384	2 solid inclusions

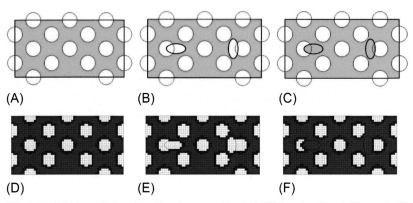

FIG. 6.65 Initial topologies of the short beam and identical FCM mesh: (A) and (D) case 1; (B) and (E) case 2; and (C) and (F) case 3.

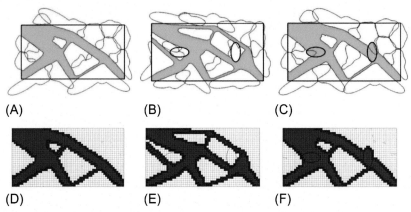

FIG. 6.66 Final designs and FCM meshes in three cases: (A) and (D) case 1; (B) and (E) case 2; and (C) and (F) case 3.

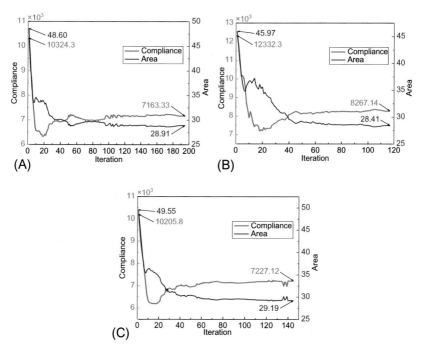

FIG. 6.67 Convergence histories of the structural compliance and volume: (A) case 1; (B) case 2; and (C) case 3.

TABLE 6.5 Several intermediate iterations for different cases.

Items	Case 1	Case 2	Case 3
Iteration 10			
Iteration 20			
Iteration 40			
Iteration 100			

In this section, a new optimization framework for freeform design domain structures is introduced. The topology optimization process is transformed as an action of the topology variation modeler (TVM) onto the freeform design domain modeler (FDDM), as modeled in Section 5.1.

6.4.1 RBF-based topology optimization

Example 6.13 Topology optimization of a cantilever beam with a hole.
The structure is shown in Fig. 6.68. This is a classical topology optimization problem of an irregular design domain that is widely studied (Sigmund, 2001; Belytschko et al., 2003; Xing et al., 2010; Zhou and Wang, 2013; Qian, 2013). Here, the design domain is divided into 90×60 cells and all the supports of CS-RBFs have a 3.5 mm radius. By using the level-set function Φ constructed in Eq. (5.17) and ignoring the stress constraints, we can obtain the optimized configuration after 116 iterations when the volume fraction is constrained to be 0.5. Intermediate design results are illustrated in Fig. 6.69B and C. The final design of this test is found to be the same as in the previous works of other scholars. Notice that the Young's modulus is 207.4 GPa and Poisson's ratio is 0.3. All dimensions are in millimeters.

In order to fully demonstrate the applicability of our optimization procedure, the design domain of this cantilever beam is further modified (see Fig. 6.70) and the hole radius is also taken as an additional design variable with the lower bound of 10 mm. The construction process of the level-set function, Φ, is schematically illustrated in Fig. 6.71. The finite cell model related to the initial design and the corresponding stress analysis result are depicted in Fig. 6.72.

First, the compliance minimizing is investigated with the volume constraint $V \le 10,000 \text{ mm}^3$. Fig. 6.73 plots the design results without stress constraint. Fig. 6.73A indicates that the radius reaches its lower bound. Fig. 6.73B indicates

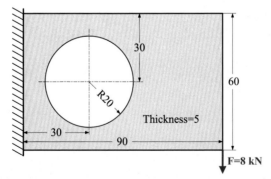

FIG. 6.68 Design domain of a cantilever beam with a fixed hole.

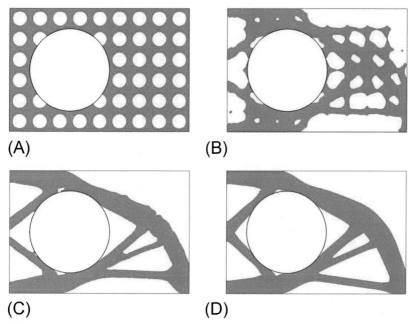

(A)

(B)

(C)

(D)

FIG. 6.69 Topology optimization of the cantilever beam with a fixed hole: (A) initial topology generated by the TVM and FDDM; (B and C) intermediate topology designs; and (D) final topology design.

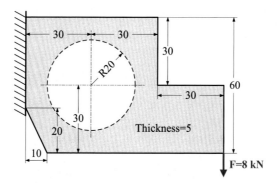

FIG. 6.70 Modified design domain of the cantilever beam with a designable hole.

that stress concentrations occur at the material corner of 270° and both ends of the fixed boundary.

Fig. 6.74 shows the optimized results subjected to stress constraints. Although the final design has the same topology as in Fig. 6.73, the local stress level is well controlled below $\bar{\sigma} = 800\,\mathrm{MPa}$ owing to the use of Eq. (5.21). With

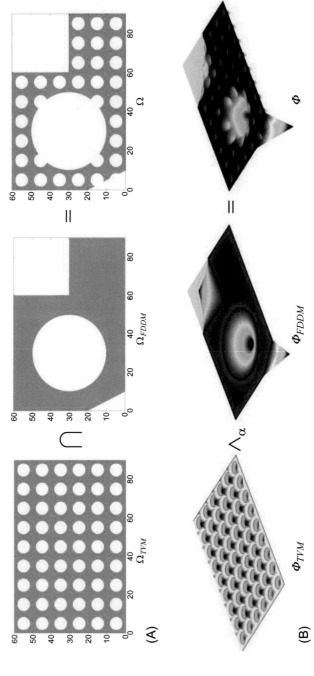

FIG. 6.71 Construction of the level-set function ϕ used in the optimization of the cantilever beam: (A) Boolean set operation of related domains; and (B) corresponding level-set functions.

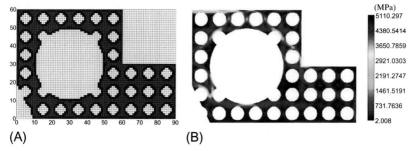

(A)   (B)

FIG. 6.72 Analysis of the initial model of the cantilever beam: (A) finite cell mesh consisting of 90×60 cells; and (B) contour plot of von Mises stress.

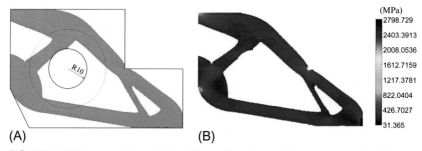

(A)   (B)

FIG. 6.73 Minimum compliance design of the cantilever beam without stress constraint: (A) zero level-set contour of the final design ($J = 2444\,\text{N mm}$, $V = 9978\,\text{mm}^3$); and (B) contour plot of von Mises stress.

the active stress constraint technique in Section 5.2.2, Fig. 6.74D shows that only $n_a/m = 2$ stress constraints are involved in each group toward the final optimization stage.

Example 6.14 Topology optimization of a two-point loading mechanical part. Fig. 6.75 shows a mechanical part including four annuli marked in black as nondesign regions. This problem belongs to design case 2 indicated in Fig. 5.8B. The level-set function Φ shown in Fig. 6.76 is composed of three level-set functions Φ_{TVM}, Φ_{FDDM}^{basic}, and Φ_{FDDM}^{aux} according to Eq. (5.17). In this test, a 160×80 mesh is used for the finite cell analysis and all the supports of CS-RBFs related to Φ_{TVM} have a radius of 3.5 mm. The FCM analysis model and stress result of the initial design are plotted in Fig. 6.77.

As in the previous test, both cases of minimizing compliance with and without stress constraint are studied here for the purpose of comparison. The upper bound of the volume is $\overline{V} = 20{,}000\,\text{mm}^3$, which is nearly half the design domain volume.

Figs. 6.78 and 6.79 show the design results without and with stress constraint, respectively. Different topologies are obtained. Especially, in the second

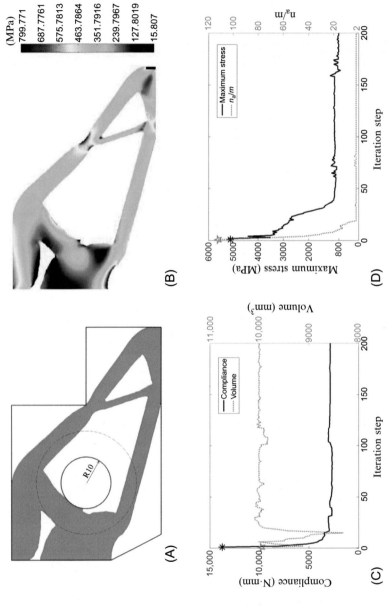

FIG. 6.74 Minimum compliance design of the cantilever beam with stress constraint: (A) zero level-set contour of the final design ($J = 2846\,\text{N}\cdot\text{mm}$, $V = 9998\,\text{mm}^3$); (B) contour plot of von Mises stress; (C) convergence histories of the compliance J and volume V; and (D) iteration histories of the maximum von Mises stress and the number of constraints clustered by each aggregation function.

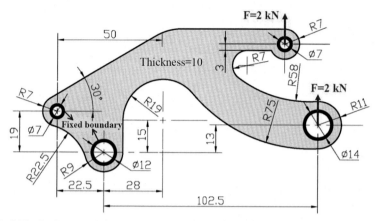

FIG. 6.75 Design domain of a two-point loading mechanical part.

case, there rests only one local constraint in each group ($n_a/m = 1$) at the last iterations so that the maximum stress is exactly controlled in the final design shown in Fig. 6.79B and D.

Example 6.15 Topology optimization of a bracket with an embedded component.

This problem is inspired from the traditional shape optimization studied in Bennett and Botkin (1985). In the design domain described in Fig. 6.80, the non-designable solid features correspond to four inner hole rings marked in black with a minimum width and should be preserved during the optimization process. According to Eq. (5.17), the corresponding level-set function, Φ, can be constructed in the way shown in Fig. 6.81. Fig. 6.82 depicts the finite cell model and stress analysis result of the initial design. The radius is set to be 7 mm for the supports of CS-RBFs in this test.

The optimization problem formulated in Eq. (5.21) is solved, and the volume is limited by 18,000 mm^3 as the upper bound, which is about 40% of the initial volume of the design domain. To show the generality, both the position and the wall thickness of the component are taken as design variables. The variation of the position is not restricted, and the change of the wall thickness is limited between 0.5 mm and 9.5 mm. Notice that the volume of this movable component is not included in the total volume of the design results. Otherwise, the component will be excluded from the optimized configuration with its wall thickness reaching the lower bound of 0.5 mm, as shown in Fig. 6.83. Because with the same volume, the material distributed freely out of the nondesign region can always gain a faster decrease of compliance than the one used to increase the wall thickness of the component.

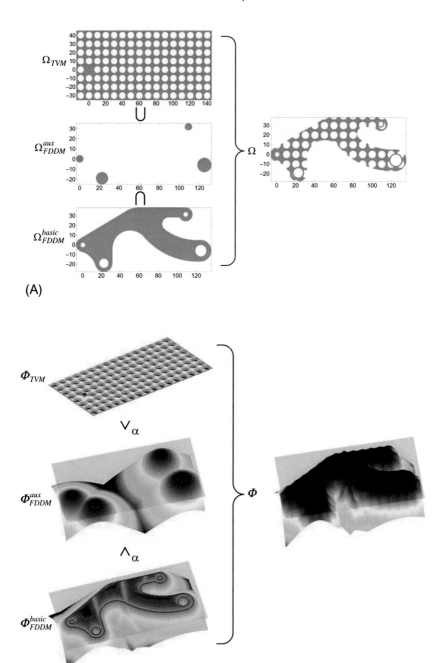

FIG. 6.76 Construction of the level-set function Φ used in the optimization of the connecting plate: (A) Boolean set operation of related domains; and (B) corresponding level-set functions.

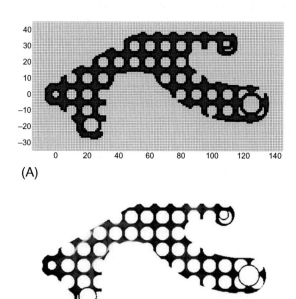

FIG. 6.77 Analysis of the initial model of the cantilever beam: (A) finite cell mesh consisting of 160×80 cells; and (B) contour plot of von Mises stress.

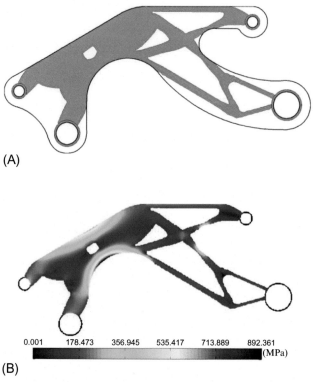

FIG. 6.78 Minimum compliance design of the mechanical part without stress constraint: (A) zero level-set contour of the final design ($J = 2773\,\text{N·mm}$, $V = 19{,}964\,\text{mm}^3$); and (B) contour plot of von Mises stress.

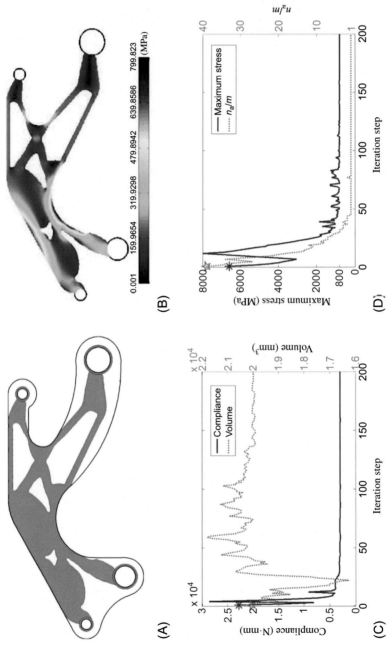

FIG. 6.79 Minimum compliance design of the mechanical part with stress constraint: (A) zero level-set contour of the final design ($J = 2979\,$N·mm, $V = 19{,}981\,$mm^3); (B) contour plot of von Mises stress; (C) convergence histories of the compliance J and volume V; and (D) iteration histories of the maximum von Mises stress and the number of constraints clustered by each aggregation function.

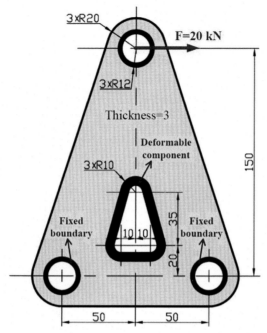

FIG. 6.80 Design domain of a bracket with an embedded triangle-like component.

Fig. 6.84 shows the final design results of the present problem. It is seen that the component acts as an essential part of the structure and its wall thickness reaches the upper bound of 9.5 mm. In addition, the local stress level is controlled precisely as only one local constraint is involved in each group at the last iterations. After about 50 iterations, the structural topology hardly changes along with the convergence of compliance and the maximum stress, as shown in Fig. 6.84D and E. Some representative design results at the early optimization stages are given in Fig. 6.85 to show clearly the topological changes.

6.4.2 Feature-driven topology optimization

Example 6.16 Topology optimization of a clipped cantilever beam with a designable hole.

The structure is shown in Fig. 6.70. Three design models with 48 solid, 58 void, and 106 mixed solid-void superellipses are tested. Initial design models and their higher-dimensional level-set functions are shown in Fig. 6.86 and Fig. 6.87. The boundaries of solid and void features are distinguished in red and green. The initial dimension of each solid superellipse is 12 mm × 2 mm. And the sizes of 3.4 mm × 3.4 mm and 2 mm × 2 mm are used for void

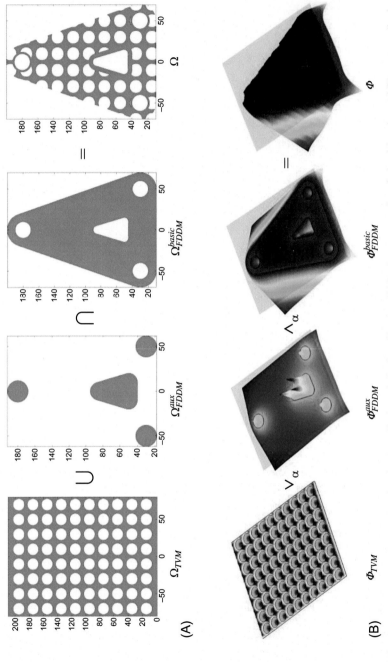

FIG. 6.81 Construction of the level-set function Φ used in the optimization of bracket: (A) Boolean set operation of related domains; and (B) corresponding level-set functions.

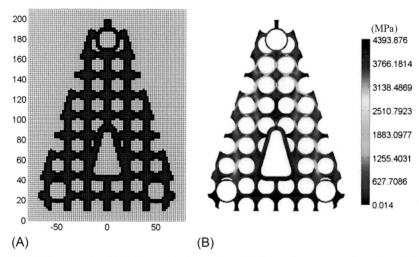

FIG. 6.82 Analysis of the initial model of the bracket: (A) finite cell mesh consisting of 80×105 cells; and (B) contour plot of von Mises stress.

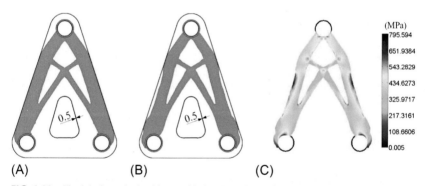

FIG. 6.83 Final designs obtained by considering the volume of the component in the volume constraint: (A) the optimized design without stress constraint; (B) the optimized design with stress constraint; and (C) its von Mises stress contour.

superellipses in design model 2 and 3, respectively. Optimized topologies are obtained after the movement, rotation, and deformation of superelliptical features and are shown in Fig. 6.88. Corresponding higher-dimensional level-set functions are illustrated in Fig. 6.89. The convergence histories of compliance and volume are shown in Fig. 6.90. All cases converge and satisfy the prescribed volume constraint. The final topologies agree well with the result obtained with the RBF-based level-set method shown in Fig. 6.73 and the hole radius reaches its lower bound in all cases. Although small differences exist, the compliances

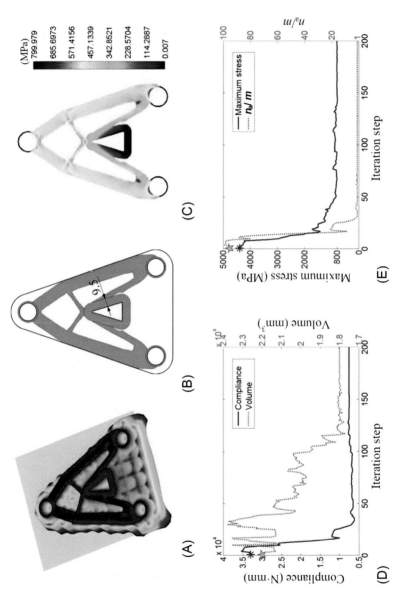

FIG. 6.84 Minimum compliance design of the bracket with stress constraint: (A) Level-set function Φ of the final design; (B) zero level-set contour ($J = 7501\,\mathrm{N\cdot mm}$, $V = 17{,}979\,\mathrm{mm}^3$); (C) contour plot of von Mises stress; (D) convergence histories of the compliance J and volume V; and (E) iteration histories of the maximum von Mises stress and the number of constraints clustered by each aggregation function.

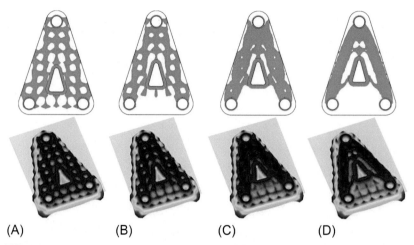

FIG. 6.85 Zero level-set contours and the corresponding level-set functions at some early stage iterations: (A) iteration 10, (B) iteration 22, (C) iteration 30, and (D) iteration 41.

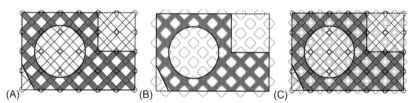

FIG. 6.86 Initial topologies of the clipped cantilever beam: (A) with pure solid features; (B) with pure void features; and (C) with mixed solid-void features.

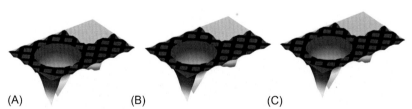

FIG. 6.87 Higher dimensional level-set functions for initial topologies of the clipped cantilever beam: (A) with pure solid features; (B) with pure void features; and (C) with mixed solid-void features.

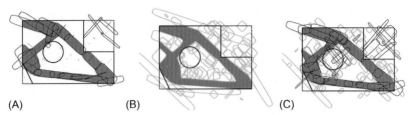

FIG. 6.88 Optimized topologies of the clipped cantilever beam: (A) with pure solid features; (B) with pure void features; and (C) with mixed solid-void features.

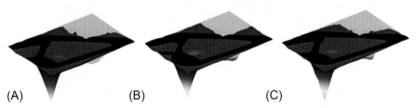

FIG. 6.89 Higher dimensional level-set functions for optimized topologies of the clipped cantilever beam: (A) with pure solid features; (B) with pure void features; and (C) with mixed solid-void features.

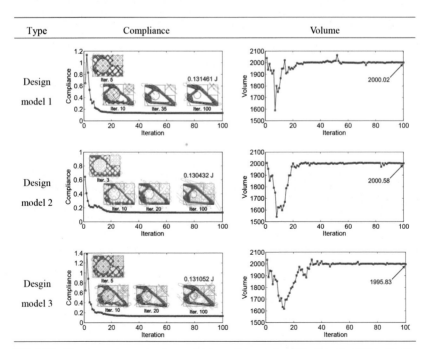

FIG. 6.90 Convergence curves and several intermediate results of the clipped cantilever beam.

are nearly the same. It is worth mentioning that thousands of design variables associated with RBF coefficients are needed in Example 6.13 while 240, 290, and 530 design variables are available in the current study.

Example 6.17 Topology optimization of a bracket with a designable triangle-like feature.

Fig. 6.80 shows the design domain containing four holed rings as intrinsic features. As in the previous test, three design models with pure solid, pure void, and mixed solid-void features are studied for the purpose of comparison. Their initial topologies contain 40 solid, 54 void, and 94 mixed solid-void superellipses, as shown in Fig. 6.91. Fig. 6.92 shows the final design results. The corresponding higher-dimensional level-set functions of the initial and optimized topologies are illustrated in Figs. 6.93 and 6.94, respectively. Convergence curves and intermediate results are depicted in Fig. 6.95. As the volume of the triangle-like feature is included in the calculation of the overall structural volume, its wall thickness reaches the lower bound of 0.5 mm in this test.

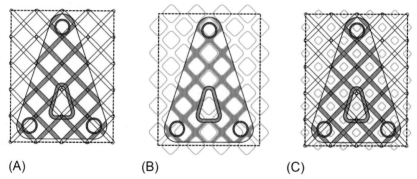

(A)   (B)   (C)

FIG. 6.91 Initial topologies of the bracket: (A) with pure solid features; (B) with pure void features; and (C) with mixed solid-void features.

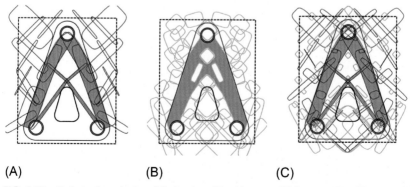

(A)   (B)   (C)

FIG. 6.92 Optimized topologies of the bracket: (A) with pure solid features; (B) with pure void features; and (C) with mixed solid-void features.

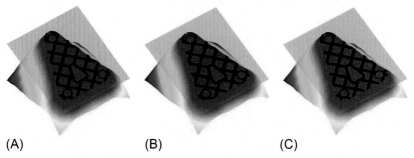

FIG. 6.93 Higher dimensional level-set functions for initial topologies of the bracket: (A) with pure solid features; (B) with pure void features; and (C) with mixed solid-void features.

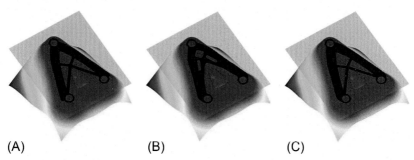

FIG. 6.94 Higher dimensional level-set functions for optimized topologies of the bracket: (A) with pure solid features; (B) with pure void features; and (C) with mixed solid-void features.

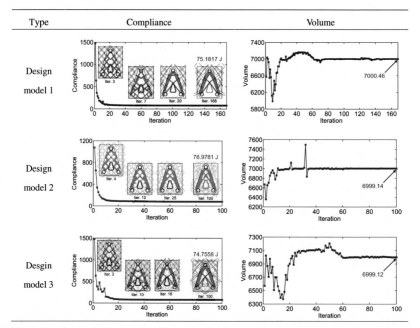

FIG. 6.95 Convergence curves and intermediate results of the bracket.

It can be seen that the optimized topologies are comparable with those obtained with the RBF-based level-set method shown in Fig. 6.83A. In detail, the inner three holes in Fig. 6.92B take the shape of a superellipse because only one void superellipse is used for the definition of each hole. A similar phenomenon exists for the definition of solid feature members connecting bottom holed rings in Fig. 6.92A. Besides, as the initial layout and shape of solid features in Figs. 6.91A and C are identical, the final designs are nearly the same. But the design model with mixed solid-void features is flexible to refine the detailed design near the holed bottom rings owing to the introduction of additional void features.

Example 6.18 Topology optimization of a periodic structure.
Fig. 6.96A shows a rectangular design domain to be topologically optimized as a periodic structure. It is defined as periodic arrays of a representative volume element (RVE) whose dimension is $l \times h$. Due to the periodicity, the design models of all RVEs are the same as that of the original RVE, to which the FDDM is directly related and the TVM is defined with solid and/or void features. In this test, the RVE has a size of 16 mm $\times$ 10 mm. A total number of $m = 2 \times 2 = 4$ RVEs then exist. A vertically distributed force F $= 100$ N/mm is applied along the right edge of the whole structure. The Young's modulus and Poisson's ratio of the solid material are $E = 1000$ MPa and $\nu = 0.3$, respectively. The ratio of "ersatz material" to solid material is 1×10^{-5} for Young's modulus. The upper bound of the volume fraction is set to be 60%, and equally a volume of 384 mm^3.

Consider an arbitrary point (x, y) in the RVE of row i and column j. The number i and j of the RVE can be calculated as

$$\begin{cases} i = [x/l + 1] \\ j = [y/h + 1] \end{cases} \tag{6.22}$$

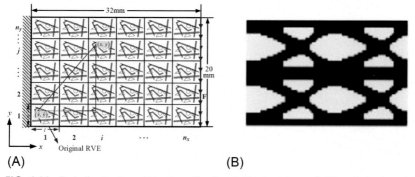

(A)   (B)

FIG. 6.96 Periodic structure: (A) schematic of periodic structure; and (B) optimized result obtained with SIMP method in Zhang and Sun (2006).

with [·] denoting the maximum integer not bigger than.

The mapping from point (x, y) to a specific point $(\widetilde{x}, \widetilde{y})$ in the original RVE can then be expressed as

$$\begin{cases} \widetilde{x} = x - (i-1)l \\ \widetilde{y} = y - (j-1)h \end{cases} \qquad (6.23)$$

Due to the periodicity, the level-set function of point (x, y) equals that of point $(\widetilde{x}, \widetilde{y})$ and the equality $\Phi(x,y) = \Phi(\widetilde{x}, \widetilde{y})$ holds.

Different initial topologies are tested to show their influences. First, three different RVEs consisting of nine solid, seven void, and 13 solid-void superellipses are considered as design models in Fig. 6.97. The whole structures are also depicted. Fig. 6.98 shows the optimized results. The solution obtained by the density method (Zhang and Sun, 2006) is shown in Fig. 6.96B for

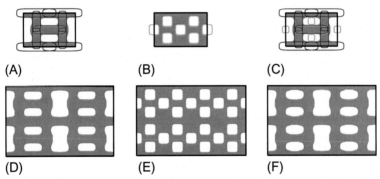

(A)   (B)   (C)

(D)   (E)   (F)

FIG. 6.97 First kind of initial topologies of RVEs and the whole structures: (A) RVE with pure solid features; (B) RVE with pure void features; (C) RVE with mixed solid-void features; (D) whole structure 1; (E) whole structure 2; and (F) whole structure 3.

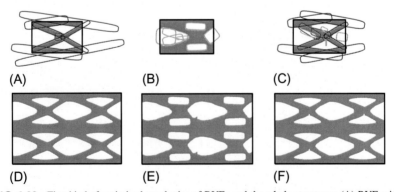

(A)   (B)   (C)

(D)   (E)   (F)

FIG. 6.98 First kind of optimized topologies of RVEs and the whole structures: (A) RVE with pure solid features; (B) RVE with pure void features; (C) RVE with mixed solid-void features; (D) whole structure 1; (E) whole structure 2; and (F) whole structure 3.

FIG. 6.99 Higher dimensional level-set functions for the first kind of initial topologies of the whole structure: (A) with pure solid features; (B) with pure void features; and (C) with mixed solid-void features.

FIG. 6.100 Higher dimensional level-set functions for the first kind of optimized topologies of the whole structure: (A) with pure solid features; (B) with pure void features; and (C) with mixed solid-void features.

comparison. The higher-dimensional level-set functions of the initial and optimized results are shown in Figs. 6.99 and 6.100. Convergence histories are shown in Fig. 6.101. As in the previous test, some inner holes in Fig. 6.98B are only defined by one superellipse. Likewise, it is reasonable that the mean compliance of the optimized structure with mixed solid-void features in Fig. 6.98C is slightly less than that with the pure solids in Fig. 6.98A.

Example 6.19 Topology optimization of a cyclic symmetry structure.
To highlight the advantage of the proposed method, the optimization of a disc problem is studied and the necessity of the unified feature definition formulation is illustrated. The disc problem shown in Fig. 6.102A is a typical cyclic symmetry structure that can be regarded as periodic arrays of original RVE along the circumferential direction, as shown in Fig. 6.102B. Suppose the structure is partitioned into n_c RVEs of identical topology. The inner and outer predefined skins of thickness t are the so-called intrinsic features. The domain encircled by the skins is topologically designable. Suppose $t = 1$ mm and $n_c = 16$ in this example. The boundary of the inner hole is clamped and a point-wise tangential force $F = 1600$ N is applied on the boundary of the outer skin. The Young's modulus and Poisson's ratio of the solid material are $E = 1000$ MPa and $\nu = 0.3$, respectively. The ratio of "ersatz material" to solid material is 1×10^{-5} for Young's modulus.

Consider an arbitrary point (x, y) in the RVE of number i_c. Its angle θ relative to the disc's center corresponds to

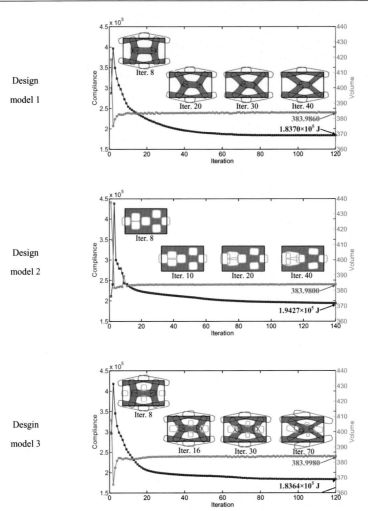

FIG. 6.101 Convergence curves and intermediate results of the bracket.

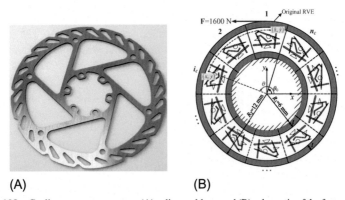

(A)   (B)

FIG. 6.102 Cyclic symmetry structure: (A) a disc problem; and (B) schematic of the feature-based design model.

$$\theta = \arctan\left(\frac{y}{x}\right) + \frac{\pi}{2}\operatorname{sign}(x)[\operatorname{sign}(x) - 1] \tag{6.24}$$

The number i_c of the RVE can be calculated as

$$i_c = \begin{cases} \left[\dfrac{\theta - \theta_0}{2\pi} \cdot n_c + 1\right] & \theta \geq \theta_0 \\[3mm] \left[\dfrac{\theta + 2\pi - \theta_0}{2\pi} \cdot n_c + 1\right] & \theta < \theta_0 \end{cases} \tag{6.25}$$

where θ_0 is the angle of the original RVE, as shown in Fig. 6.102B.

The mapping from point (x, y) to a specific point $(\widetilde{x}, \widetilde{y})$ in the original RVE can then be expressed as

$$\begin{cases} \widetilde{x} = \cos\left(\theta - \dfrac{2\pi}{n_c}(i_c - 1)\right) \\[3mm] \widetilde{y} = \sin\left(\theta - \dfrac{2\pi}{n_c}(i_c - 1)\right) \end{cases} \tag{6.26}$$

Due to the cyclic symmetry property, the level-set function of point (x, y) equals that of point $(\widetilde{x}, \widetilde{y})$ with $\Phi(x, y) = \Phi(\widetilde{x}, \widetilde{y})$.

First, design models defined with pure solid and pure void features are tested. We reduce the constrained volume bound gradually. The optimized structures are shown in Figs. 6.103 and 6.104, respectively. Note that the x-axis depicts the prescribed volume bound while the y-axes denote the corresponding compliances and volumes of optimized structures. It is observed that when the upper bound of the volume is extremely small (such as when the constrained

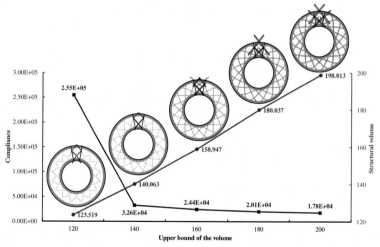

FIG. 6.103   Optimized structures defined with pure solid features for different volume bounds.

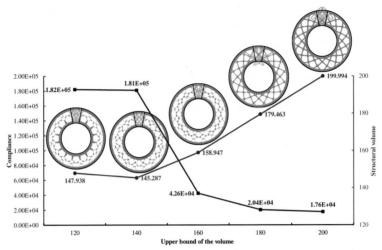

FIG. 6.104 Optimized structures defined with pure void features for different volume bounds.

volume equals 120 or 140), the optimized structures obtained with pure solid or pure void features are disconnected. Besides, the volume of optimized structure with pure void features is much larger than the constrained volume bound.

In order to satisfy the extremely small volume constraint, void features are further set and limited within the intrinsic feature domain with the finite circle method (Zhu et al., 2010). Thus, two design models related to the mixed solid-void features and void-void features are applied, respectively. The optimized structures are depicted in Figs. 6.105 and 6.106, respectively. It is shown that

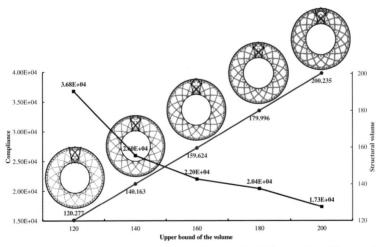

FIG. 6.105 Optimized structures defined with mixed solid-void features for different volume bounds.

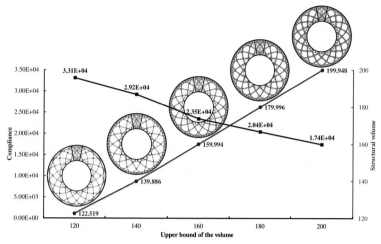

FIG. 6.106 Optimized structures defined with pure void features (void features are designable within the intrinsic feature domain) for different volume bounds.

the introduction of void features within intrinsic features is beneficial in satisfying prescribed volume constraint and obtaining connected structures.

6.4.3   CBS-based topology optimization

Example 6.20 CBSs-driven topology optimization of a torque arm.

This problem originates from a benchmark of shape optimization and is now considered a topology optimization problem. To show the generality, two solid rings with a width of 1 cm considered as intrinsic features are preserved during the optimization process. CBSs are used as basic design primitives for construction of TVM.

Two cases listed in Table 6.6 are considered to reveal the influence of the number of design variables upon the topology optimization. Fig. 6.107A and B illustrate the torque arm model with initial layouts of CBS. Fig. 6.107C and D

TABLE 6.6 Two cases of CBS for the torque arm.

Items	Number of CBSs	Number of design variables for each CBS	Number of total design variables
Case 1	11	24	264
Case 2	25	24	600

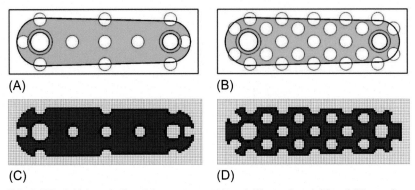

FIG. 6.107 Initial topologies of the torque arm: (A) and (C) case 1; and (B) and (D) case 2.

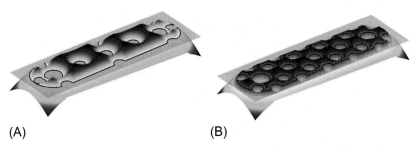

FIG. 6.108 LSFs of the initial topologies for the torque arm: (A) case 1; and (B) case 2.

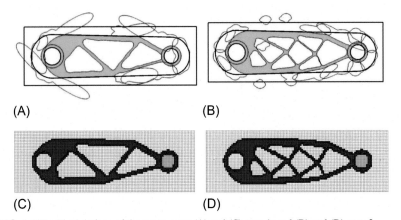

FIG. 6.109 Final designs of the torque arm: (A) and (C) case 1; and (B) and (D) case 2.

show the analysis model embedded into a regular rectangle domain that is discretized into a regular FCM mesh of 122×40. Fig. 6.108 shows the initial level-set function of the torque arm.

Fig. 6.109 shows the optimized topologies outlined by the CBSs together with the FCM mesh. Fig. 6.110 shows the level-set functions of the final designs

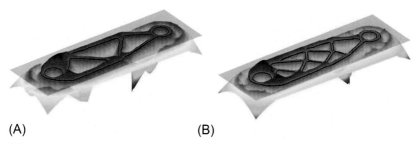

FIG. 6.110   LSFs of the final designs for the torque arm: (A) case 1; and (B) case 2.

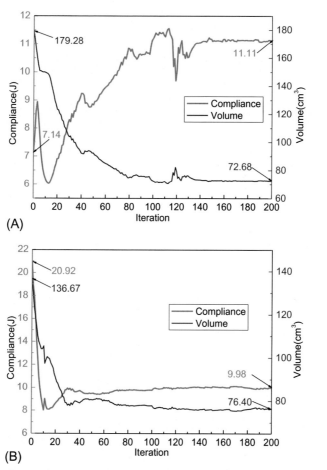

FIG. 6.111   Convergence histories of the structural compliance and volume for the torque arm: (A) case 1; and (B) case 2.

TABLE 6.7 Intermediate results of the torque arm in both cases.

Items	Case 1 11 CBS, 264 Design variables	Case 2 25 CBS, 600 Design variables
Iteration 10		
Iteration 20		
Iteration 40		
Iteration 100		

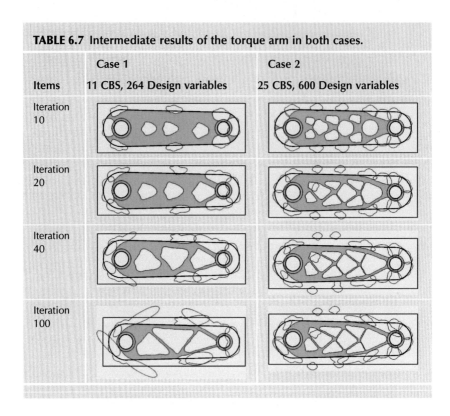

for the torque arm. Fig. 6.111 gives the convergence histories of structural compliance and volume. In view of the number of design variables, Case 2 yields a topology of enriched holes that ameliorate further the structural compliance than in Case 1. Intermediate results are given in Table 6.7 to show the stable evolutions of structural topologies in both cases. Nondesignable solid features are found to be well preserved by Boolean operations during the optimization process.

References

Alexandrov, O., Santosa, F., 2005. A topology-preserving level set method for shape optimization. J. Comput. Phys. 204 (1), 121–130.

Allaire, G., Jouve, F., Toader, A.M., 2004. Structural optimization using sensitivity analysis and a level-set method. J. Comput. Phys. 194 (1), 363–393.

Belytschko, T., Xiao, S.P., Parimi, C., 2003. Topology optimization with implicit functions and regularization. Int. J. Numer. Methods Eng. 57 (8), 1177–1196.

Bendsøe, M.P., Kikuchi, N., 1988. Generating optimal topologies in structural design using a homogenization method. Comput. Methods Appl. Mech. Eng. 71, 197–224.

Bendsøe, M.P., Sigmund, O., 1999. Material interpolation schemes in topology optimization. Arch. Appl. Mech. 69 (9–10), 635–654.

Bennett, J.A., Botkin, M.E., 1985. Structural shape optimization with geometric description and adaptive mesh refinement. AIAA J. 23 (3), 458–464.

Bletzinger, K.U., Firl, M., Linhard, J., Wüchner, R., 2010. Optimal shapes of mechanically motivated surfaces. Comput. Methods Appl. Mech. Eng. 199 (5–8), 324–333.

Boresi, A.P., Schmidt, R.J., Sidebottom, O.M., 1985. Advanced Mechanics of Materials. Vol. 6. Wiley, New York.

Braibant, V., Fleury, C., 1984. Shape optimal design using B-splines. Comput. Methods Appl. Mech. Eng. 44 (3), 247–267.

Buhl, T., 2002. Simultaneous topology optimization of structure and supports. Struct. Multidiscip. Optim. 23 (5), 336–346.

Burger, M., Hackl, B., Ring, W., 2004. Incorporating topological derivatives into level set methods. J. Comput. Phys. 194 (1), 344–362.

Chen, J., Shapiro, V., Suresh, K., Tsukanov, I., 2007. Shape optimization with topological changes and parametric control. Int. J. Numer. Methods Eng. 71 (3), 313–346.

Deaton, J.D., Grandhi, R.V., 2014. A survey of structural and multidisciplinary continuum topology optimization: post 2000. Struct. Multidiscip. Optim. 49 (1), 1–38.

Dunning, P.D., Kim, H.A., Mullineux, G., 2011. Investigation and improvement of sensitivity computation using the area-fraction weighted fixed grid FEM and structural optimization. Finite Elem. Anal. Des. 47 (8), 933–941.

Düster, A., Parvizian, J., Yang, Z., Rank, E., 2008. The finite cell method for three-dimensional problems of solid mechanics. Comput. Methods Appl. Mech. Eng. 197 (45–48), 3768–3782.

Fougerolle, Y.D., Gribok, A., Foufou, S., Truchetet, F., Abidi, M.A., 2005. Boolean operations with implicit and parametric representation of primitives using R-functions. IEEE Trans. Vis. Comput. Graph. 11 (5), 529–539.

Gao, H.H., Zhu, J.H., Zhang, W.H., Zhou, Y., 2015. An improved adaptive constraint aggregation for integrated layout and topology optimization. Comput. Methods Appl. Mech. Eng. 289, 387–408.

García-Ruíz, M.J., Steven, G.P., 1999a. Fixed Grid Finite Element Analysis in Structural Design and Optimization (Doctoral Dissertation, PhD Thesis). The University of Sydney, Sydney. Referenced in 71.

García-Ruíz, M.J., Steven, G.P., 1999b. Fixed grid finite elements in elasticity problems. Eng. Comput. 16, 145–164.

Guest, J.K., Prévost, J.H., Belytschko, T., 2004. Achieving minimum length scale in topology optimization using nodal design variables and projection functions. Int. J. Numer. Methods Eng. 61 (2), 238–254.

Haslinger, J., Mäkinen, R.A., 2003. Introduction to Shape Optimization: Theory, Approximation, and Computation. Society for Industrial and Applied Mathematics.

James, K.A., Martins, J.R., 2012. An isoparametric approach to level set topology optimization using a body-fitted finite-element mesh. Comput. Struct. 90, 97–106.

James, K.A., Lee, E., Martins, J.R., 2012. Stress-based topology optimization using an isoparametric level set method. Finite Elem. Anal. Des. 58, 20–30.

Kang, Z., Wang, Y., 2013. Integrated topology optimization with embedded movable holes based on combined description by material density and level sets. Comput. Methods Appl. Mech. Eng. 255, 1–13.

Kim, N.H., Chang, Y., 2005. Eulerian shape design sensitivity analysis and optimization with a fixed grid. Comput. Methods Appl. Mech. Eng. 194 (30–33), 3291–3314.

Kim, N.H., Choi, K.K., Botkin, M.E., 2002. Numerical method for shape optimization using mesh-free method. Struct. Multidiscip. Optim. 24 (6), 418–429.

Kumar, A.V., Padmanabhan, S., Burla, R., 2008. Implicit boundary method for finite element analysis using non-conforming mesh or grid. Int. J. Numer. Methods Eng. 74 (9), 1421–1447.

Laurent-Gengoux, P., Mekhilef, M., 1993. Optimization of a NURBS representation. Comput. Aided Des. 25 (11), 699–710.

Le, C., Norato, J., Bruns, T., Ha, C., Tortorelli, D., 2010. Stress-based topology optimization for continua. Struct. Multidiscip. Optim. 41 (4), 605–620.

Le, C., Bruns, T., Tortorelli, D., 2011. A gradient-based, parameter-free approach to shape optimization. Comput. Methods Appl. Mech. Eng. 200 (9–12), 985–996.

Liu, J., Ma, Y.S., 2015. 3D level-set topology optimization: a machining feature-based approach. Struct. Multidiscip. Optim. 52 (3), 563–582.

Luo, Z., Wang, M.Y., Wang, S., Wei, P., 2008. A level set-based parameterization method for structural shape and topology optimization. Int. J. Numer. Methods Eng. 76 (1), 1–26.

Luo, Z., Zhang, N., Gao, W., Ma, H., 2012. Structural shape and topology optimization using a meshless Galerkin level set method. Int. J. Numer. Methods Eng. 90 (3), 369–389.

Mei, Y., Wang, X., Cheng, G., 2008. A feature-based topological optimization for structure design. Adv. Eng. Softw. 39 (2), 71–87.

Norato, J.A., Bendsøe, M.P., Haber, R.B., Tortorelli, D.A., 2007. A topological derivative method for topology optimization. Struct. Multidiscip. Optim. 33 (4–5), 375–386.

Norato, J.A., Bell, B.K., Tortorelli, D.A., 2015. A geometry projection method for continuum-based topology optimization with discrete elements. Comput. Methods Appl. Mech. Eng. 293, 306–327.

Olhoff, N., 1988. Minimum stiffness of optimally located supports for maximum value of beam eigenfrequencies. J. Sound Vib. 120, 457–463.

Osher, S., Sethian, J.A., 1988. Fronts propagating with curvature-dependent speed: algorithms based on Hamilton-Jacobi formulations. J. Comput. Phys. 79 (1), 12–49.

Parvizian, J., Düster, A., Rank, E., 2007. Finite cell method. Comput. Mech. 41 (1), 121–133.

Parvizian, J., Düster, A., Rank, E., 2012. Topology optimization using the finite cell method. Optim. Eng. 13 (1), 57–78.

Qian, X., 2013. Topology optimization in B-spline space. Comput. Methods Appl. Mech. Eng. 265, 15–35.

Qian, Z., Ananthasuresh, G.K., 2004. Optimal embedding of rigid objects in the topology design of structures. Mech. Based Des. Struct. Mach. 32 (2), 165–193.

Radovcic, Y., Remouchamps, A., 2002. BOSS QUATTRO: an open system for parametric design. Struct. Multidiscip. Optim. 23 (2), 140–152.

Rall, L.B., 1981. Automatic Differentiation: Techniques and Applications. Lect. Notes Comput. Sci., 120Springer Verlag, Berlin, Germany.

Rozvany, G., 1974. Optimization of unspecified generalized forces in structural design. J. Appl. Mech., Trans. ASME 41 (4), 1143–1145.

Sethian, J.A., Wiegmann, A., 2000. Structural boundary design via level set and immersed interface methods. J. Comput. Phys. 163 (2), 489–528.

Shapiro, V., 1991. Theory of R-Functions and Applications: A Primer. Cornell University.

Shapiro, V., 2007. Semi-analytic geometry with R-functions. ACTA Numer. 16, 239–303.

Sigmund, O., 2001. A 99 line topology optimization code written in Matlab. Struct. Multidiscip. Optim. 21 (2), 120–127.

Sigmund, O., Maute, K., 2013. Topology optimization approaches. Struct. Multidiscip. Optim. 48 (6), 1031–1055.

Stolpe, M., Svanberg, K., 2001. An alternative interpolation scheme for minimum compliance topology optimization. Struct. Multidiscip. Optim. 22 (2), 116–124.

Sukumar, N., Moës, N., Moran, B., Belytschko, T., 2000. Extended finite element method for three-dimensional crack modeling. Int. J. Numer. Methods Eng. 48 (11), 1549–1570.

Svanberg, K., 1995. A globally convergent version of MMA without linesearch. In: Proceedings of the First World Congress of Structural and Multidisciplinary Optimization, Goslar, Germany. Vol. 28, pp. 9–16.

Tanskanen, P., 2002. The evolutionary structural optimization method: theoretical aspects. Comput. Methods Appl. Mech. Eng. 191 (47–48), 5485–5498.

Van Miegroet, L., 2012. Generalized Shape Optimization Using XFEM and Level Set Description. (Doctoral Dissertation). Université de Liège, Liège, Belgique.

Van Miegroet, L., Duysinx, P., 2007. Stress concentration minimization of 2D filets using X-FEM and level set description. Struct. Multidiscip. Optim. 33 (4–5), 425–438.

Wang, M.Y., Li, L., 2013. Shape equilibrium constraint: a strategy for stress-constrained structural topology optimization. Struct. Multidiscip. Optim. 47 (3), 335–352.

Wang, S., Wang, M.Y., 2006. Radial basis functions and level set method for structural topology optimization. Int. J. Numer. Methods Eng. 65 (12), 2060–2090.

Wang, D., Zhang, W., 2013. A general material perturbation method using fixed mesh for stress sensitivity analysis and structural shape optimization. Comput. Struct. 129, 40–53.

Wang, M.Y., Wang, X., Guo, D., 2003. A level set method for structural topology optimization. Comput. Methods Appl. Mech. Eng. 192 (1–2), 227–246.

Wang, F., Lazarov, B.S., Sigmund, O., 2011. On projection methods, convergence and robust formulations in topology optimization. Struct. Multidiscip. Optim. 43 (6), 767–784.

Wang, Y., Luo, Z., Kang, Z., Zhang, N., 2015. A multi-material level set-based topology and shape optimization method. Comput. Methods Appl. Mech. Eng. 283, 1570–1586.

Xia, L., Zhu, J., Zhang, W., 2012. Sensitivity analysis with the modified Heaviside function for the optimal layout design of multi-component systems. Comput. Methods Appl. Mech. Eng. 241, 142–154.

Xia, L., Zhu, J., Zhang, W., Breitkopf, P., 2013. An implicit model for the integrated optimization of component layout and structure topology. Comput. Methods Appl. Mech. Eng. 257, 87–102.

Xia, Q., Wang, M.Y., Shi, T., 2014. A level set method for shape and topology optimization of both structure and support of continuum structures. Comput. Methods Appl. Mech. Eng. 272, 340–353.

Xie, Y.M., Steven, G.P., 1997. Basic evolutionary structural optimization. In: Evolutionary Structural Optimization. Springer, London, pp. 12–29.

Xing, X., Wei, P., Wang, M.Y., 2010. A finite element-based level set method for structural optimization. Int. J. Numer. Methods Eng. 82 (7), 805–842.

Zhang, W.H., Sun, S., 2006. Scale-related topology optimization of cellular materials and structures. Int. J. Numer. Methods Eng. 68 (9), 993–1011.

Zhang, W.H., Beckers, P., Fleury, C., 1995. A unified parametric design approach to structural shape optimization. Int. J. Numer. Methods Eng. 38 (13), 2283–2292.

Zhang, W., Wang, D., Yang, J., 2010. A parametric mapping method for curve shape optimization on 3D panel structures. Int. J. Numer. Methods Eng. 84 (4), 485–504.

Zhang, W., Xia, L., Zhu, J., Zhang, Q., 2011. Some recent advances in the integrated layout design of multicomponent systems. J. Mech. Des. 133(10).

Zhang, J., Zhang, W.H., Zhu, J.H., Xia, L., 2012. Integrated layout design of multi-component systems using XFEM and analytical sensitivity analysis. Comput. Methods Appl. Mech. Eng. 245, 75–89.

Zhou, M., Wang, M.Y., 2013. Engineering feature design for level set based structural optimization. Comput. Aided Des. 45 (12), 1524–1537.

Zhu, J.H., Zhang, W.H., 2006. Maximization of structural natural frequency with optimal support layout. Struct. Multidiscip. Optim. 31 (6), 462–469.

Zhu, J.H., Zhang, W.H., 2010. Integrated layout design of supports and structures. Comput. Methods Appl. Mech. Eng. 199 (9–12), 557–569.

Zhu, J.H., Beckers, P., Zhang, W.H., 2010. On the multi-component layout design with inertial force. J. Comput. Appl. Math. 234 (7), 2222–2230.

Zhu, J.H., Zhang, W.H., Xia, L., 2016. Topology optimization in aircraft and aerospace structures design. Arch. Comput. Meth. Eng. 23 (4), 595–622.

Chapter 7

Feature-driven optimization for structures under design-dependent loads

7.1 Topology optimization including design-dependent body loads

Generally, a design-dependent body load concerns self-weight, the inertial load with acceleration, the thermal load in a structure, or the centrifugal load of a rotating part. It can be recognized as a particular load case following the rule of removing material resulting in removing load.

In the framework of the density-based method, the main concern is how to control the parasitic effect of the low density region in the presence of design-dependent body loads. Rodrigues and Fernandes (1995) adopted the homogenization method to study the design-dependent temperature load effect on the optimized design in the earlier time. Turteltaub and Washabaugh (1999) investigated the influences of combined centrifugal and external loads upon the optimal material distribution of a rotating prismatic structure. As design variables refer to pseudodensities that are directly attached to the FE model of a structure to drive topology variation, material interpolation models such as SIMP (Bendsøe and Sigmund, 2003; Sigmund and Maute, 2013) and RAMP (Stolpe and Svanberg, 2001) were improved to penalize the element property and load in terms of pseudodensity variables. The self-weight problem was initially studied by Bruyneel and Duysinx (2005). It was found that the presence of the low-density region is due to the mismatched penalization between the stiffness and load. Meanwhile, the structural compliance is nonmonotonous with respect to the pseudodensity variable and the volume constraint may be inactive at the optimum solution. They then exploited a modified SIMP model with a linear penalization at low density and also suggested the RAMP model.

Later, the polynomial interpolation model was developed by Zhu et al. (2009) to balance the penalty between the design-dependent body load and the stiffness. Gao and Zhang (2010) and Gao et al. (2016) illustrated that the RAMP model is superior to the SIMP model in the presence of thermoelastic

The Feature-driven Method for Structural Optimization. https://doi.org/10.1016/B978-0-12-821330-8.00007-9

stress loads. Xu et al. (2013) utilized an optimality criteria method to compare SIMP and RAMP models for topology optimization, including self-weight or centrifugal force, and also found that RAMP is convenient to such problems.

Evolutionary structural optimization (ESO) and bi-directional evolutionary structural optimization (BESO) can be regarded as discrete versions of the density-based method with density variables taking 0–1 discrete values. This method was adopted by Yang et al. (2005) in the case of self-weight and supporting pressure. It was found that BESO has the flexibility of balancing the solution quality and computing time. Ansola et al. (2006) presented a modified version of ESO in computing sensitivity numbers to obtain the optimum design for the self-weight problems. Huang and Xie (2011) developed a BESO method to deal with the self-weight problems using the RAMP model. The level-set method is an alternative to solve this kind of problem. Xia and Wang (2008) made remarkable progress by developing the level-set method of the discrete form to circumvent low-density regions caused by the standard density-based method. Recently, Xia and Shi (2016) introduced the level-set-based multiple-type boundary method to realize simultaneous topology optimization in terms of different types of boundaries.

In this section, design-dependent body loads are dealt with by means of the closed B-splines (CBS) presented in Chapter 2. Each CBS serves as a deformable hole in the design domain to realize topological changes. As design variables are the control parameters of the CBS boundary shape and location, traditional material interpolation models are no longer needed to artificially penalize the material properties, so the number of design variables could be greatly reduced. As a result, topology optimization is realized in the way of shape optimization to produce optimized topologies free of low-density regions and jagged boundaries. It is seen that the implementation can easily be achieved because the CBS can be represented in both forms of parametric and implicit equations. The parametric form is beneficial for the problem formulation with a small number of design variables while the implicit form acting as the level-set function is beneficial both for structural reanalysis using the fixed grid of the FCM (Parvizian et al., 2007) and for the topology description with Boolean operations.

7.1.1 CBS-based model for topology optimization

Eq. (4.8) concerns the weak-form equilibrium equation of 2D linear thermoelasticity. Here, the body force f and the thermal stress load σ^{th} are considered design-dependent loads. They are strongly dependent upon the material layout whenever topology optimization is carried out over domain Ω. Typical body forces and thermal stress loads are given as

$$f = \begin{cases} \rho g, & \text{self} - \text{weight load} \\ \rho a, & \text{acceleration load} \\ \rho \omega^2 X, & \text{centrifugal force} \end{cases} \tag{7.1}$$

and

$$\boldsymbol{\sigma}^{\text{th}} = \boldsymbol{C} : \boldsymbol{\varepsilon}^{\text{th}} = \boldsymbol{C} : \gamma \Delta T \boldsymbol{I} \tag{7.2}$$

in which $\boldsymbol{g}$, $\boldsymbol{a}$, and ω denote the gravitational acceleration, acceleration, and angular velocity, respectively. ρ and γ denote the material density and thermal expansion coefficient. $\boldsymbol{\sigma}^{\text{th}}$ is the thermal stress tensor induced by the temperature rise ΔT. $\boldsymbol{C}$ is the elastic constitutive tensor. In the case of 2D plane stress problems, $\boldsymbol{I} = [1,1,0]^T$ implies that the temperature rise does not influence the shear strain.

In the framework of the density-based method, the body force depends upon the pseudodensity variable of each element to model the effect of design-dependence. Material interpolation models are then established on the basis of the FE model. Basically, there exist four representative interpolation models: the standard SIMP model (Bendsøe and Sigmund, 2003; Sigmund and Maute, 2013), the modified SIMP model (Bruyneel and Duysinx, 2005), the RAMP model (Stolpe and Svanberg, 2001; Zhang et al., 2016), and the polynomial interpolation model (Zhu et al., 2009; Zhang et al., 2016), as summarized in Table 7.1. Here, p, q and w are the prescribed parameters.

Fig. 7.1 shows the corresponding penalty functions and ratio functions of these models. Notice that the last two models have a smooth penalty function as well as a positive slope at $\eta_i = 0$.

In the framework of the level-set method, the topology of an elastic body is represented by the level-set function $\Phi(\boldsymbol{x})$, whose control parameters are then considered design variables to drive boundary variations. Thus, both the body force and thermal stress load can be decoupled with design variables so that the total potential energy of the concerned structure is expressed as

$$\Pi = \frac{1}{2} \int_{\text{D}} H(\Phi) \boldsymbol{\sigma}^T \boldsymbol{\varepsilon} d\Omega - \int_{\text{D}} H(\Phi) \boldsymbol{u}^T \boldsymbol{f} d\Omega - \int_{\Gamma_N} \boldsymbol{u}^T \boldsymbol{t} d\Gamma - \int_{\text{D}} H(\Phi) \left(\boldsymbol{\sigma}^{\text{th}} \right)^T \boldsymbol{\varepsilon} d\Omega$$

$$\tag{7.3}$$

with subscript D denoting the embedding domain and $H(\cdot)$ being the Heaviside function. Clearly, the density-based method and the level-set method constitute two different formulations of design variables in dealing with design-dependent body loads.

In the CBS-based model, suppose the topology of a structure is described by Boolean operations of a set of CBSs related to level-set functions $\{\phi_i\}$. Then, the level-set function $\Phi(\boldsymbol{x})$ of the constructed topology variation modeler corresponds to

$$\Phi(\boldsymbol{x}) = - \overset{m}{\underset{i=1}{\vee}} \phi_i \tag{7.4}$$

where the negative sign implies that each CBS represents a hole, and m denotes the number of CBSs. It can be seen that topology optimization is realized in the way of shape optimization of CBS curves.

TABLE 7.1 Typical material interpolation models.

Models	Equations	Penalty function	Properties
Standard SIMP (Bendsøe and Sigmund, 2003; Sigmund and Maute, 2013; Zhu et al., 2016; Zhang et al., 2016)	$\begin{cases} \rho_i = \eta_i \rho_{i0} \\ E_i = P(\eta_i) E_{i0} \end{cases}$	$P(\eta_i) = \eta_i^p$	Zero slope at $\eta_i = 0$
Modified SIMP (Bruyneel and Duysinx, 2005)		$P(\eta_i) = \begin{cases} \eta_i^p & \eta_C < \eta_i \\ \eta_i \eta_C^{p-1} & \eta_i \le \eta_C \end{cases}$	Positive slope at $\eta_i = 0$; nondifferentiable at $\eta_i = \eta_C$
RAMP (Zhang et al., 2016; Stolpe and Svanberg, 2001)		$P(\eta_i) = \frac{\eta_i}{1 + q(1-\eta_i)}$	Positive slope at $\eta_i = 0$; differentiable everywhere
Polynomial interpolation (Zhang et al., 2016; Zhu et al., 2009)		$P(\eta_i) = (1-w)\eta_i^p + w\eta_i$	Adjustable slope at $\eta_i = 0$

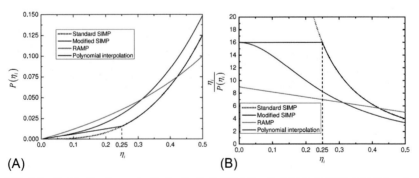

(A)

(B)

FIG. 7.1 Penalty function and ratio functions of different material interpolation models ($p = 3$, $\eta_C = 0.25$, $q = 8$, $w = 1/16$): (A) penalty functions; and (B) ratios of pseudodensity variable to the penalty function.

7.1.2 Sensitivity analysis of the CBS-based model including design-dependent body loads

In this section, the topology optimization problem concerns the minimization of structural compliance. As is known, the sensitivity of the displacement vector U can be derived from Eq. (4.28)

$$\frac{\partial U}{\partial d_i} = K^{-1}\left(\frac{\partial F}{\partial d_i} - \frac{\partial K}{\partial d_i}U\right) \tag{7.5}$$

Consequently, the sensitivity of structural compliance J with respect to design variable d_i is calculated as

$$\frac{\partial J}{\partial d_i} = \frac{\partial F^T}{\partial d_i}U + F^T\frac{\partial U}{\partial d_i} = 2\frac{\partial F^T}{\partial d_i}U - U^T\frac{\partial K}{\partial d_i}U \tag{7.6}$$

To proceed, consider the following general integral form

$$\Psi(x, d) = \int_D \psi(x)H(\Phi(x, d))d\Omega \tag{7.7}$$

The sensitivity of the above function can be calculated as

$$\frac{\partial \Psi(x, d)}{\partial d_i} = \int_D \psi(x)\frac{\partial H(\Phi(x, d))}{\partial \Phi(x, d)}\frac{\partial \Phi(x, d)}{\partial d_i}d\Omega = \int_{\partial\Omega} \psi(x)\frac{\partial \Phi}{\partial d_i}\frac{1}{\|\nabla\Phi\|}d\Gamma \tag{7.8}$$

with

$$\|\nabla\Phi\| = \sqrt{\left(\frac{\partial \Phi}{\partial x}\right)^2 + \left(\frac{\partial \Phi}{\partial y}\right)^2} \tag{7.9}$$

When the variety of design-dependent loads given below are considered, derivatives of the load vector in Eq. (7.6) can easily be obtained by Eq. (7.8)

$$\psi(x) = \begin{cases} 1, & \text{structural volume} \\ B^T CB, & \text{stiffness matrix} \\ M^T\rho g, & \text{self} - \text{weight load} \\ M^T\rho a, & \text{acceleration load} \\ M^T\rho\omega^2 X, & \text{centrifugal force} \\ B^T\sigma^{\text{th}}, & \text{thermal stress load} \end{cases} \tag{7.10}$$

Numerically, $\partial\Phi/\partial d_i$ and $\|\nabla\Phi\|$ in Eq. (7.8) are computed with respect to the Gaussian points along the physical boundary. According to Eq. (7.4), we can obviously obtain the following derivative relations.

$$
\begin{cases}
\dfrac{\partial \Phi(x,d)}{\partial d_i} = \dfrac{\partial \Phi(x,d)}{\partial \phi_i} \cdot \dfrac{\partial \phi_i(x,d_i)}{\partial d_i} \\[4mm]
\|\nabla \Phi(x,d)\| = \left\| \displaystyle\sum_{i=1}^{n} \dfrac{\partial \Phi(x,d)}{\partial \phi_i} \nabla \phi_i(x,d_i) \right\|
\end{cases}
\tag{7.11}
$$

As d_i only concerns $\phi_i(x,d_i)$ of each CBS, computations can approximately be simplified as

$$
\begin{cases}
\dfrac{\partial \Phi(x,d)}{\partial d_i} = -\dfrac{\partial \phi_i(x,d_i)}{\partial d_i} \\[4mm]
\|\nabla \Phi(x,d)\| = \|\nabla \phi_i(x,d_i)\|
\end{cases}
\tag{7.12}
$$

Here, design variables of each CBS consist of the center coordinates (x_0, y_0) and the control parameters $R_1, R_2, \ldots, R_n$, namely, $d_i = \{x_0, y_0, R_1, R_2, \ldots, R_n\}$. To clarify this idea, consider two primitives $\phi_1(x,d_1)$ and $\phi_2(x,d_2)$ as an example. By resorting to the R-function operator, we can write

$$
\Phi(x,d) = -\phi_1 \vee \phi_2 = -\left(\phi_1 + \phi_2 + \sqrt{\phi_1^2 + \phi_2^2} \right)
\tag{7.13}
$$

and

$$
\frac{\partial \Phi(x,d)}{\partial \phi_i} = -\left(1 + \frac{\phi_i}{\sqrt{\phi_1^2 + \phi_2^2}} \right), \quad i = 1,2
\tag{7.14}
$$

Especially at the boundary $\phi_i = 0$, the above equation can be simplified as $\partial \Phi(x,d)/\partial \Phi_i = -1$. It is then used in Eq. (7.11) for sensitivity analysis. Here, Eq. (7.15) is adopted for the representation of each CBS.

$$
\phi(x,y,r) = r^2 - (x - x_0)^2 - (y - y_0)^2
\tag{7.15}
$$

In fact, the expression of the level-set function is indifferent for sensitivity analysis when the boundary integral scheme of Eq. (5.42) is used. The corresponding derivative with respect to R_i reads

$$
\frac{\partial \phi}{\partial R_i} = \frac{\partial \phi}{\partial r} \frac{\partial r}{\partial R_i} = 2r \cdot B_{i,p}\left(\frac{\theta + \pi/2}{2\pi} \right)
\tag{7.16}
$$

The partial derivatives with respect to coordinates x and y correspond to

$$
\frac{\partial \phi}{\partial x} = -\frac{\partial \phi}{\partial x_0} = 2r \cdot \sum_{i=1}^{n} R_i \frac{\partial B_{i,p}}{\partial \theta} \frac{\partial \theta}{\partial x} - 2(x - x_0)
\tag{7.17}
$$

$$
\frac{\partial \phi}{\partial y} = -\frac{\partial \phi}{\partial y_0} = 2r \cdot \sum_{i=1}^{n} R_i \frac{\partial B_{i,p}}{\partial \theta} \frac{\partial \theta}{\partial y} - 2(y - y_0)
\tag{7.18}
$$

According to Eq. (2.57), the partial derivatives of angle θ are calculated as

$$\begin{cases} \dfrac{\partial \theta}{\partial x} = -\dfrac{y - y_0}{(x - x_0)^2 + (y - y_0)^2} \\[4mm] \dfrac{\partial \theta}{\partial y} = \dfrac{x - x_0}{(x - x_0)^2 + (y - y_0)^2} \end{cases} \qquad (7.19)$$

7.1.3 Numerical examples including design-dependent body loads

Several numerical examples are studied to validate the proposed topology optimization method. The implementation of the globally convergent method of moving asymptotes (GCMMA) (Svanberg, 1995) within the Boss-Quattro optimization platform (Radovcic and Remouchamps, 2002) is used as the optimizer.

Example 7.1 Topology optimization of the arch structure involving self-weight
This example was extensively studied as a benchmark by several researchers (Bruyneel and Duysinx, 2005; Yang et al., 2005; Huang and Xie, 2011). The Young's modulus and Poisson's ratio are assumed to be $E = 1$ and $\mu = 0.3$, respectively. As shown in Fig. 7.2, a vertical force is applied at the middle point of the structure bottom. The height of the structure is $L = 6$ and the gravitational acceleration is set to $g = 10$. The problem is to minimize the structural compliance with the volume fraction bounded by 30%. The structure is discretized into a 40×20 FCM mesh. During the optimization process, boundary cells are instantly identified and refined using the quadtree approach to ensure the accuracy of Gaussian integration.

First, suppose only the self-weight exists while the vertical force is omitted. The inequality and equality constraints of the structural volume are considered as two study cases. Fig. 7.3 depicts the initial topology and corresponding FCM mesh. Seventeen CBSs with 14 design variables for each of them are available to realize the topological changes. Fig. 7.4 shows the optimized topologies indicating that different volume constraints have great effects on the optimized

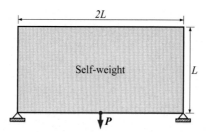

FIG. 7.2 The arch structure with self-weight.

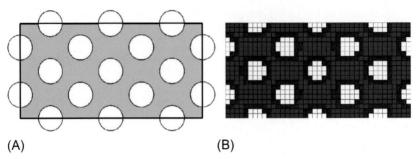

(A)   (B)

FIG. 7.3 Initial topology and corresponding FCM mesh with only self-weight: (A) initial topology; and (B) initial FCM mesh.

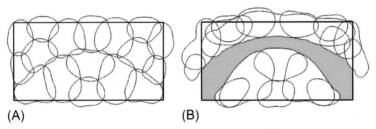

(A)   (B)

FIG. 7.4 Optimized topologies with only self-weight subjected to inequality and equality volume constraints: (A) volume fraction is less than 30%; and (B) volume fraction equals 30%.

results. Particularly, the inequality volume constraint yields a void structure, which is obviously consistent with the common sense that no materials lead to no self-weight and then no structure. Fig. 7.5 shows the FCM mesh of optimized topology for the equality volume constraint. Fig. 7.6 gives the convergence histories of the compliance and volume.

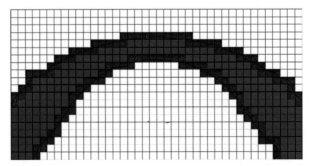

FIG. 7.5 FCM mesh of optimized topology for equality volume constraint.

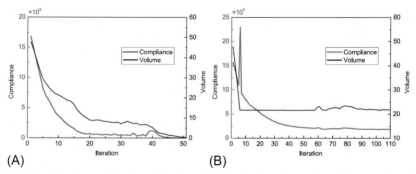

FIG. 7.6 Convergence histories with only self-weight for inequality and equality constraints of structural volume: (A) volume fraction is less than 30%; and (B) volume fraction equals 30%.

Second, the self-weight and vertical force are simultaneously applied to the structure. Meanwhile, different values of the vertical force are tested to illustrate their influences on the optimized topologies. Similarly to the aforementioned initial layout, 16 CBSs with 14 design variables for each of them are now adopted, as shown in Fig. 7.7. Optimized topologies for different load cases are nearly the same, as illustrated in Fig. 7.8. Corresponding FCM meshes are given in Fig. 7.9.

Clearly, when P is small, much material is used near the supports to resist the self-weight. When P is large, much material migrates near the concentrated force to resist the traction. Convergence histories of compliance and volume for different load cases are given in Fig. 7.10. At the beginning of the iterations, the compliance increases sharply and then decreases smoothly due to the violation of volume constraint.

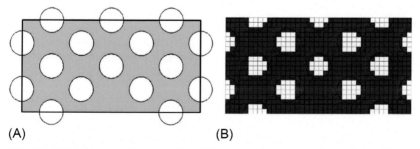

FIG. 7.7 Initial topology and corresponding FCM mesh with self-weight and concentrated force: (A) initial topology; and (B) initial FCM mesh.

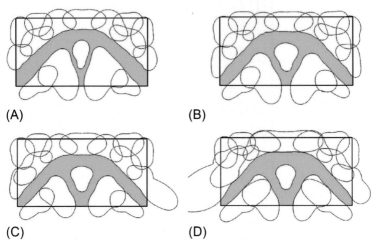

FIG. 7.8   Optimized topologies for different load cases: (A) $P = 100$ with self-weight; (B) $P = 500$ with self-weight; (C) $P = 1000$ with self-weight; and (D) $P = 1000$ without self-weight.

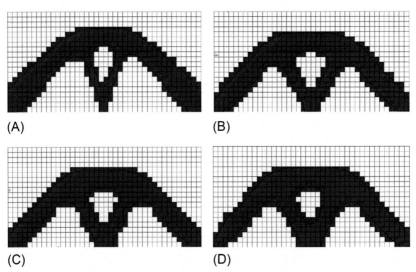

FIG. 7.9   Optimized topologies and related FCM meshes for different load cases: (A) $P = 100$ with self-weight; (B) $P = 500$ with self-weight; (C) $P = 1000$ with self-weight; and (D) $P = 1000$ without self-weight.

Example 7.2 Topology optimization of a bridge-like structure involving self-weight and nondesignable domain

Fig. 7.11 depicts the bridge-like structure with dimension parameters $L = 6$ and $t = 0.5$. Material properties are the same as in the previous example. A nondesignable domain is placed at the top. The compliance is minimized with an

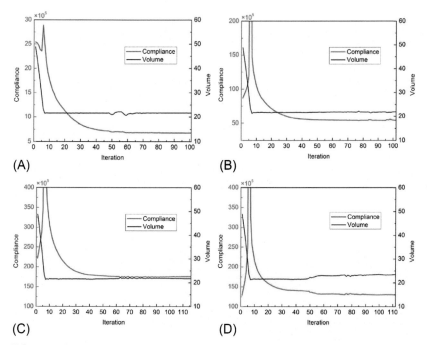

FIG. 7.10 Convergence histories of compliance and volume for different load cases: (A) $P = 100$ with self-weight; (B) $P = 500$ with self-weight; (C) $P = 1000$ with self-weight; and (D) $P = 1000$ without self-weight.

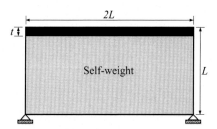

FIG. 7.11 The schematic of the bridge-like structure.

upper bound of 50% for the volume fraction. The problem is discretized with a 60×30 FCM mesh.

The initial and optimized designs of this structure are given in Fig. 7.12. Seventeen CBSs with 14 design variables for each of them are used to construct the initial layout. The corresponding FCM meshes are shown in Fig. 7.13. Convergence histories of the compliance and volume are shown in Fig. 7.14 with relevant values listed in Table 7.2. It is worth noting that the volume constraint in this problem is inactive and the final volume fraction is much smaller than 50% of the

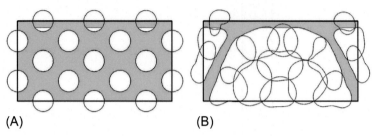

FIG. 7.12 Initial and optimized designs of the bridge-like structure: (A) initial design; and (B) optimized design.

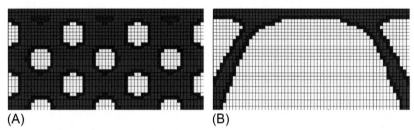

FIG. 7.13 Initial and optimized FCM meshes of the bridge-like structure: (A) initial FCM mesh; and (B) optimized FCM mesh.

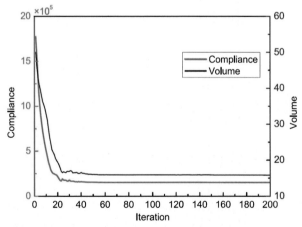

FIG. 7.14 Convergence histories of the compliance and volume.

TABLE 7.2 Compliance and volume results of initial and optimized designs.

	Compliance	Volume
Initial design	1,775,710	50.07800
Optimized design	152,602.0	15.86530

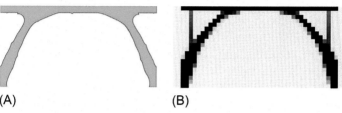

(A)   (B)

FIG. 7.15 Comparison of the optimized results for the bridge-like structure: (A) feature-driven method; and (B) Bruyneel and Duysinx (Bruyneel and Duysinx, 2005).

design domain. The final compliance reduces sharply and is only about 10% of the initial compliance. Compared to the SIMP solution (Bruyneel and Duysinx, 2005) depicted in Fig. 7.15B, the result is distinct and free of low-density elements.

Example 7.3 Topology optimization of a bi-clamped rectangular structure involving thermal stress load

This example was originally investigated by Rodrigues and Fernandes (1995) using the homogenization method; it was also studied by Gao and Zhang (2010) using the SIMP and RAMP methods. Now, the proposed method in this section is exploited as expected. As depicted in Fig. 7.16, the rectangular structure is clamped at both sides and subjected to a vertical load 10 kN at the middle bottom. Two nondesignable domains marked in dark with a width of 2.4 cm are preserved during the optimization process. The Young's modulus, Poisson's ratio, and thermal expansion coefficient of the solid material are $E = 210 \text{GPa}$, $\mu = 0.3$, and $\gamma = 1.1 \times 10^{-5}/°C$. This structure has dimensions of $72 \text{cm} \times 47.7 \text{cm} \times 1 \text{cm}$. Here, the volume is limited by the upper bound 1373.76cm^3, which is exactly 40% of the total volume.

Four cases with different temperature rises in the physical domain are considered to illustrate the influences of the thermal stress loads on the optimized designs. Fig. 7.17 shows the initial configuration and the corresponding FCM mesh of the structure. It is seen that there are 16 CBSs with 224 design variables in the initial layout. Each CBS with 14 design variables can deform freely to realize topological changes. The structure is uniformly discretized into a 60×40 FCM mesh.

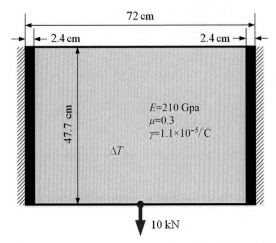

FIG. 7.16 Schematic of the bi-clamped rectangular structure (Rodrigues and Fernandes, 1995).

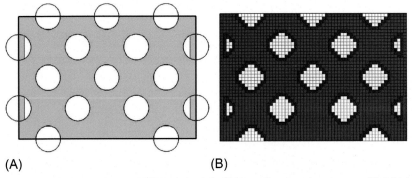

(A)   (B)

FIG. 7.17 Initial topology and FCM mesh of the bi-clamped rectangular structure: (A) initial topology; and (B) initial FCM mesh.

Fig. 7.18 shows the optimized topologies for different temperature rises and Fig. 7.19 gives the corresponding FCM meshes. Obviously, the temperature rises have great impact on the optimized results. The skeletons of the structure tend toward slender with the increased temperature rise. Fig. 7.20 depicts the convergence curves of the structural compliance and volume for different temperature rises. The initial and optimized designs are detailed in Table 7.3. Notice that the volume constraint is inactive, that is, the upper bound of the volume is not reached at the high temperature rise, which means that the corresponding optimized configurations are more advantageous than an entirely solid structure. To compare, some related results in Rodrigues and Fernandes (1995) and Gao and Zhang (2010) are illustrated in Fig. 7.21 to show the presence of low-density regions and the jagged boundaries in the structure. In those cases, the parameters of the SIMP and RAMP models influence the optimized results greatly.

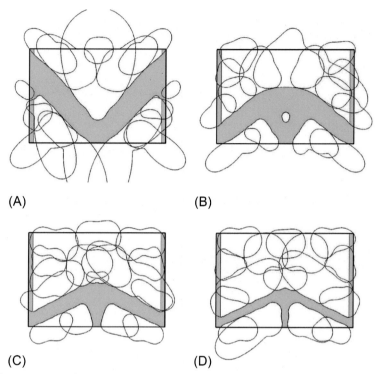

FIG. 7.18 Optimized topologies for different temperature rises: (A) $\Delta T = 0°C$; (B) $\Delta T = 1°C$; (C) $\Delta T = 4°C$; and (D) $\Delta T = 7°C$.

7.2 Concurrent shape and topology optimization involving design-dependent pressure loads

Design-dependent pressure loads are characteristic of the dependence of its location, magnitude, and direction upon the pressure boundary shape. The imposition of pressure loads should follow the moving pressure boundary and match the topology evolution in the design process. Typical cases include snow/wind loads on civil engineering structures, air pressure loads on vessels and aircraft fuselages/wings, hydrostatic pressure loads on underwater vehicles, etc.

A common solution to topology optimization of structures subjected to design-dependent pressure loads is the density-based method (Bendsøe and Sigmund, 1999; Stolpe and Svanberg, 2001). Within this framework, two different approaches are developed to deal with the pressure loads applied on designable boundaries. The first one is to interpolate the level-set contour of pseudodensity variables as the pressure boundary for load application by means of parametric curves. Hammer and Olhoff (2000) parameterized the isodensity

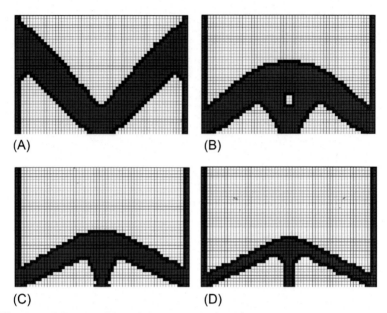

FIG. 7.19 FCM meshes of the optimized topologies for different temperature rises: (A) $\Delta T = 0°C$; (B) $\Delta T = 1°C$; (C) $\Delta T = 4°C$; and (D) $\Delta T = 7°C$.

curve from the resulting density distribution using the Bézier curve. Du and Olhoff (2004a,b) and Lee and Martins (2012) connected the isodensity points with linear segments and proposed numerical strategies to eliminate the separated or isolated boundaries that appeared in Du and Olhoff (2004a). Zhang et al. (2008) proposed an efficient boundary search scheme along the boundary of solid elements. Wang et al. (2016) regarded the intermediate topologies of SIMP as gray-scale images by which the material boundaries are identified as the zero-contour of a level-set function with the image segmentation technique. Fuchs and Shemesh (2004) predefined an independent loading surface using the Bézier curve and simultaneously optimized the shape of the pressure boundary and the topology related to density distribution.

Alternatively, pressure loads are applied by mimicking the effects of multiphysics fields. Chen and Kikuchi (2001) simulated the pressure loads with fictitious thermal loads generated by the mismatch of thermal expansion coefficients on both sides of the pressure boundary. Similar studies were extended to fluid and electric fields, etc. (Bourdin and Chambolle, 2003; Sigmund and Clausen, 2007; Bruggi and Cinquini, 2009; Zheng et al., 2009). Nevertheless, the accuracy of the pressure load application cannot be guaranteed due to the existence of ambiguous intermediate materials in the density-

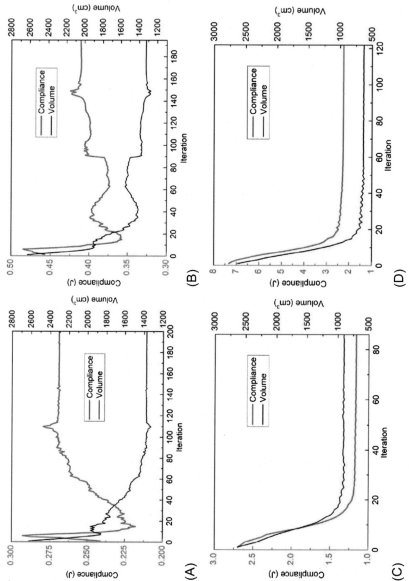

FIG. 7.20 Convergence histories of the structural compliance and volume for different temperature rises: (A) $\Delta T = 0°C$; (B) $\Delta T = 1°C$; (C) $\Delta T = 4°C$; and (D) $\Delta T = 7°C$.

TABLE 7.3 Compliance and volume results for different temperature rises.

Temperature rise ΔT (°C)	Initial design		Optimized design	
	Compliance (J)	Volume (cm³)	Compliance (J)	Volume (cm³)
0	0.241425	2622.74	0.269577	1373.72
1	0.456876	2622.74	0.410970	1325.39
4	2.69961	2622.74	1.16291	903.240
7	7.33691	2622.74	2.25460	627.610

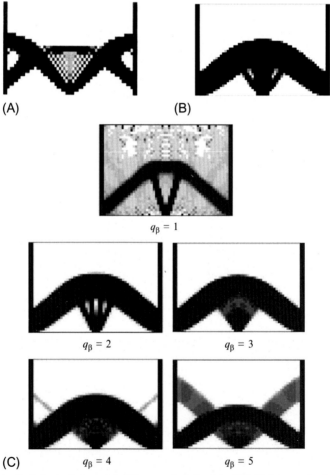

(A)   (B)

$q_\beta = 1$

$q_\beta = 2$ $q_\beta = 3$

(C) $q_\beta = 4$ $q_\beta = 5$

FIG. 7.21 Optimized solutions with different methods for $\Delta T = 1$°C: (A) homogenization method (Rodrigues and Fernandes, 1995); (B) RAMP model (Gao and Zhang, 2010); and (C) SIMP model with different parameters (Gao and Zhang, 2010).

based method. Also, the resulting structural boundaries are inevitably zig-zag because of their intrinsic discrete property.

An alternative approach is based on the so-called generalized shape optimization using the discrete level-set method (Wang et al., 2003; Allaire et al., 2004). The pressure loads applied on moving boundaries are converted into body forces within a narrow-band domain around the boundary by means of the divergence theorem (Allaire et al., 2004; Guo et al., 2004). Perturbations of hydrostatic pressure on holes and inclusions inserted by the concept of the topological derivative are considered in Xavier and Novotny (2017). Yamada et al. (2011) considered the design-dependent effect of heat convection with a discrete level-set model. Xia et al. (2015) defined the moving pressure boundary and inner topology boundaries separately with two discrete level-set functions. The avoidance of intersection between both boundaries was realized by manually modifying their velocity fields. However, the continuity of the pressure boundary cannot be exactly ensured because the discrete level-set method is a freeform optimization method with possible segmentations of the pressure boundary. Besides, derivations of the velocity field on both boundaries are mathematically intricate and limit the practical application.

Although parametric curves have been applied to describe the pressure boundary, the inner structural topologies in previous works are abstracted into mechanical models of finite element analysis. A huge number of topology design variables based on finite element or finite difference models are inevitable within the framework of traditional topology optimization methods.

A feature-based method is developed for topology optimization by which features are used as design primitives whose movements and deformations drive the evolution of structural topology. B-splines have been introduced for topology optimization in previous works (Kumar and Parthasarathy, 2011; Qian, 2013), where the density field describing the structural topology is parameterized by smooth B-spline approximations to benefit the elimination of checkerboard patterns and boundary irregularities. Parametric B-spline curves were used in the bubble method (Eschenauer et al., 1994) and the isogeometric topology optimization method (Seo et al., 2010) to achieve topology variation by adjusting the locations of control points. The optimization efficiency is largely limited due to the parametric B-spline description and the body-fitted mesh.

In this section, the feature-driven optimization method is extended for concurrent shape and topology optimization involving design-dependent pressure loads. The moving pressure boundary and inner topology boundaries of a structure are separately described using the implicit B-spline curves. To avoid the intersection between both sets of boundaries, a nondesignable offset domain of small bandwidth is artificially defined along the pressure boundary to ensure the physical imposition of the pressure load. The Neumann boundary is tractable by Boolean intersection between the moving pressure boundary and the design domain boundary. Meanwhile, the shape and topology variations of the structure are interpreted as Boolean operations in terms of level-set functions. The weighted B-spline finite cell method is applied for structural analysis.

As the fixed mesh is used, the pressure load related to the Neumann boundary is numerically discretized onto the control nodes of the finite cell method. In this section, a regularized optimization problem with perimeter regularization is solved. Sensitivities analyses of design-dependent pressure loads with respect to both shape and topology design variables are explicitly derived and calculated with the boundary integral scheme along the structural boundaries.

7.2.1 Implicit B-spline based model for concurrent shape and topology optimization

In this section, design-dependent pressure loads are considered for the traction τ on Γ_N. To be specific, the shape of the pressure boundary changes during the optimization process so that the location, magnitude, and direction of the pressure load should follow the moving pressure boundary. Meanwhile, the structure Ω is topologically optimized for the achievement of high mechanical performance and weight reduction. In this sense, the concurrent shape and topology optimization is needed.

Fig. 7.22 depicts typical pressure loads that are mathematically stated as

$$\tau = \begin{cases} c \cdot [0-1]^T & \text{snow load} \\ c \cdot \boldsymbol{n} & \text{air pressure} \\ \rho g h \cdot \boldsymbol{n} & \text{hydrostatic pressure} \end{cases} \tag{7.20}$$

in which c is a constant depending upon the magnitude of the snow load or air pressure. ρ denotes the fluid density. g is the gravitational acceleration. h refers to the vertical height between the structure and a prescribed zero reference surface of pressure. In general, the free surface of the liquid is regarded as the zero reference surface of pressure.

Here, the moving pressure boundary and inner topology boundaries are separately represented with two level-set functions. The primary concern will be

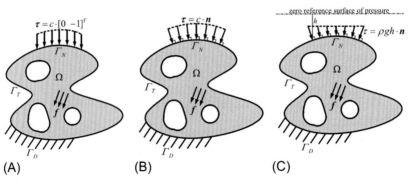

FIG. 7.22 Typical pressure loads: (A) snow load; (B) air pressure; and (C) hydrostatic pressure.

addressed on how both boundaries are described in the way of implicit B-spline curves.

(1) Representation of a moving pressure boundary

The implicit B-splines are chosen as basic primitives for the description of the moving pressure boundary in accordance with its openness and closeness. However, it is not enough to model the loading boundary with open or closed B-splines only. Below is a schematic example to illustrate this issue.

As shown in Fig. 7.23, an OBS curve is used to define an open moving boundary with the solid domain of the structure lying on its right side. Suppose both the left and bottom boundaries of the structure undergo uniform pressure load. The loading boundary is then composed of three parts: (1) the OBS curve inside the design domain; (2) the upper part of the left design domain boundary cut by the OBS curve; and (3) the right part of the bottom design domain boundary cut by the OBS curve. In this sense, the loading boundary, that is, the Neumann boundary Γ_N, can be described by means of a Boolean intersection process, as illustrated in Fig. 7.24

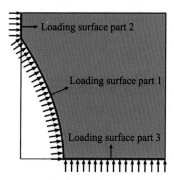

FIG. 7.23 Imposition of the pressure load on the composed Neumann boundary Γ_N.

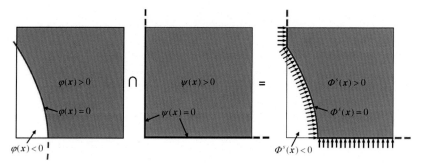

FIG. 7.24 Boolean intersection process for the accurate imposition of the pressure load on the composed Neumann boundary Γ_N.

$$\Phi^s(\boldsymbol{x}, \boldsymbol{d}^s) = \varphi(\boldsymbol{x}, \boldsymbol{d}^s) \cap \psi(\boldsymbol{x}) = \min(\varphi, \psi) = 0 \qquad (7.21)$$

where φ is the level-set function of the OBS curve as in Eq. (2.58). ψ denotes the level-set function of the design domain boundaries. In Fig. 7.24, the level-set functions of the left and bottom design domain boundaries can be stated as $x=0$ and $y=0$, respectively. We thus have $\psi(x,y)=\min(x,y)$. Notice that $\boldsymbol{d}^s = \{R_j\}$ is the vector of the shape design variables controlling the moving pressure boundary shape. Superscript s is short for "shape."

(2) Representation of inner topology boundaries

Fig. 7.25A shows the inner topology boundaries defined with a set of CBSs. Suppose the level-set function of each closed B-spline is expressed as

$$\varphi_i^t(x, y) = \sum_{j=1}^{n_i} R_{ij} B_{j,p} \left(\frac{\theta + \pi/2}{2\pi} \right) - \sqrt{(x - x_i)^2 + (y - y_i)^2} \qquad (7.22)$$

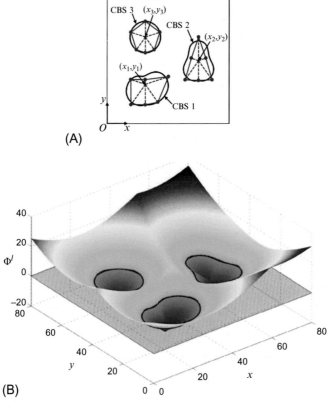

FIG. 7.25 Representation of inner topology boundaries: (A) a set of CBSs; and (B) the higher-dimensional level-set function.

where n_i corresponds to the number of control points of the ith closed B-spline. Superscript t represents the topology boundary.

Consequently, the whole level-set function of the inner topology boundaries can be obtained by means of the Boolean union

$$\Phi^t(\boldsymbol{x}, \boldsymbol{d}^t) = - \bigvee_{i=1}^{m} \varphi_i^t(\boldsymbol{x}, \boldsymbol{d}_i^t) = - \max_{i=1,\ldots,m} \left(\varphi_i^t \right) \tag{7.23}$$

where m denotes the number of closed B-splines. Here, the negative sign implies that each CBS represents a hole rather than a solid inclusion, as illustrated in Fig. 7.25B. Notice that the maximum/minimum function can mathematically be replaced in different ways with the R-function, KS function, and Ricci function. $\boldsymbol{d}^t = \{\boldsymbol{d}_i^t\}$ denotes the vector of the topology design variables to drive the topology evolution with $\boldsymbol{d}_i^t = \{x_i, y_i, R_{ij}\}$.

(3) Definition of a nondesignable offset domain

As illustrated in Fig. 7.26A, the pressure load will be wrongly applied on void materials if the pressure boundary intersects with the topology boundaries. In the previous work (Xia et al., 2015), Xia et al. avoided the intersection between both boundaries by picking up the portions close to each other in every iteration and manually forcing the normal velocity fields to keep the consistence of their evolutions in the next iteration. Nevertheless, intersections may occur, especially when the time step is large.

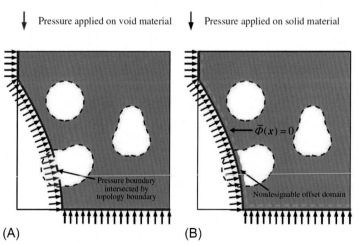

(A)   (B)

FIG. 7.26   Definition of nondesignable offset domain of small bandwidth: (A) inaccurate imposition of pressure loads on void materials at the intersection part between pressure and topology boundaries; and (B) accurate imposition of pressure loads on solid materials by introducing the nondesignable offset domain along the pressure boundary.

To overcome this difficulty, we introduce a nondesignable offset domain of small bandwidth along the pressure boundary so that the pressure load can be properly imposed on solid materials, as shown in Fig. 7.26B. The level-set function of the nondesignable offset domain can be simply given on the basis of $\Phi^s(x, d^s)$

$$\overline{\Phi}(x, d^s) = -\Phi^s(x, d^s) + \Delta \tag{7.24}$$

where Δ is the bandwidth of the nondesignable offset domain. Thus, the structural boundaries are divided into two sets of disjoint boundaries, namely the Neumann boundary $\Gamma_N = \{x \in D \mid \Phi^s(x) = 0\}$ and the traction-free boundary $\Gamma_T = \{x \in D \mid \Psi(x) = \max(\Phi^t, \overline{\Phi}) = 0\}$.

The solid domain of the structure then holds the following relation

$$\Phi(x, d^s, d^t) = \Psi(x, d^s, d^t) \wedge \Phi^s(x, d^s) = \min(\Psi, \Phi^s) \geq 0 \tag{7.25}$$

Notice that any modification of the shape design variables in d^s will drive the variation of the pressure loading boundary. Meanwhile, any change of the topology design variables in d^t controls the inner topology evolution. This is just the mechanism of the concurrent shape and topology optimization.

7.2.2 Sensitivity analysis of implicit B-spline-based model involving design-dependent pressure loads

Here, the pressure load is distributed on control points by numerical integration on the Neumann boundary Γ_N approximated by line segments. The end points of each line segment are determined by the bi-section method. With the negligence of body force f, F then corresponds to the pressure load vector. Numerically, the latter is calculated as

$$F = \int_{\Gamma_N} \overline{M}^T \tau d\Gamma = \sum_{c \in \Lambda} \sum_{i=1}^{n_c} w_i \overline{M}(x_i)^T \tau |J_c| \tag{7.26}$$

where Λ is the set of cells cutting by the Neumann boundary Γ_N. n_c denotes the number of integral points in each boundary cell. x_i and w_i refer to the coordinate of the ith integral point and its weight. $|J_c|$ denotes the determinant of the Jacobian matrix J_c for the cth cell.

The minimum compliance problem with the prescribed volume constraint is considered for the concurrent shape and topology optimization. It is commonly recognized that the above optimization problem is ill-posed so that a nonconvergent design sequence can be obtained by introducing an infinite number of microscopic holes into the structure (Bendsøe and Sigmund, 2003). Well-posed problems can be obtained by relaxation or regularization methods. The homogenization method and the density-based method are known as primary relaxation methods allowing intermediate materials (Bendsøe and Kikuchi, 1988; Bendsøe and Sigmund, 1999). Alternatively, geometrical constraints can be used to regularize the ill-posed optimization problem by adding an additional term in the original problem, such as perimeter regularization (Wei and Wang, 2009), Tikhonov regularization (Yamada et al., 2010), etc.

In this section, a regularized objective function $\bar{J} = J + \beta L(\Omega)$ is used with $L(\Omega) = \int_{\partial\Omega} d\Gamma$ and β corresponding to the structural perimeter and regularization factor, respectively. It will be shown in Section 7.2.3 that one such regularization is capable of removing unnecessary holes caused by the local optimum phenomenon.

The sensitivity of $\bar{J}$ with respect to any design variable d_i can be calculated as

$$\frac{\partial \bar{J}}{\partial d_i} = \frac{\partial J}{\partial d_i} + \beta \frac{\partial L}{\partial d_i} \tag{7.27}$$

where

$$\frac{\partial J}{\partial d_i} = \frac{1}{2} U^T \frac{\partial K}{\partial d_i} U + U^T K \frac{\partial U}{\partial d_i} = U^T \frac{\partial F}{\partial d_i} - \frac{1}{2} U^T \frac{\partial K}{\partial d_i} U \tag{7.28}$$

Compared to problems with a fixed pressure boundary, the difficulty of sensitivity analysis lies in the calculation of the derivative of the pressure load F, which is dependent on the shape of the moving pressure boundary in each iteration. It will be shown that the sensitivity of the pressure load in Eq. (7.26) can be explicitly calculated on the Neumann boundary.

To proceed, consider the following two general integral forms

$$\Psi_1(x, d) = \int_D \psi_1(x) H(\Phi(x, d)) d\Omega \tag{7.29}$$

$$\Psi_2(x, d) = \int_\Gamma \psi_2(x) d\Gamma \tag{7.30}$$

in which Γ could be either the structural boundary $\partial\Omega$ or its partial segment (e.g., Γ_N) related to the level-set function $\Phi(x,d)$.

The sensitivities of the above functions can be obtained according to relevant works (Wang et al., 2003; Allaire et al., 2004)

$$\frac{\partial \Psi_1(x)}{\partial d_i} = \int_D \psi_1(x) \frac{\partial H(\Phi)}{\partial \Phi} \frac{\partial \Phi}{\partial d_i} d\Omega = \int_{\partial\Omega} \psi_1(x) \frac{\partial \Phi}{\partial d_i} \frac{1}{\|\nabla \Phi\|} d\Gamma \tag{7.31}$$

$$\frac{\partial \Psi_2(x)}{\partial d_i} = -\int_\Gamma (\nabla \psi_2 \cdot n + \kappa \psi_2) \frac{\partial \Phi}{\partial d_i} \frac{1}{\|\nabla \Phi\|} d\Gamma \tag{7.32}$$

where ∇ refers to the gradient operator and $\|\ \|$ denotes the vector module. κ is the curvature of Γ. Notice that the above expressions are demonstrated to be independent of the mathematical expressions of the level-set function $\Phi(x,d)$.

The sensitivities of K, F, V, and L can then be obtained in accordance with Eqs. (7.31), (7.32)

$$
\begin{cases}
\dfrac{\partial K}{\partial d_i} = \displaystyle\int_{\partial\Omega} [\boldsymbol{L\overline{M}}]^T \boldsymbol{C} [\boldsymbol{L\overline{M}}] \dfrac{\partial \Phi}{\partial d_i} \dfrac{1}{\|\nabla\Phi\|} d\Gamma \\[3mm]
\dfrac{\partial V}{\partial d_i} = \displaystyle\int_{\partial\Omega} \dfrac{\partial \Phi}{\partial d_i} \dfrac{1}{\|\nabla\Phi\|} d\Gamma \\[3mm]
\dfrac{\partial F}{\partial d_i} = -\displaystyle\int_{\Gamma_N} (\nabla\overline{M}\cdot\boldsymbol{n} + \kappa\overline{M})\tau \dfrac{\partial \Phi^s}{\partial d_i} \dfrac{1}{\|\nabla\Phi^s\|} d\Gamma \\[3mm]
\dfrac{\partial L}{\partial d_i} = -\displaystyle\int_{\partial\Omega} \kappa \dfrac{\partial \Phi}{\partial d_i} \dfrac{1}{\|\nabla\Phi\|} d\Gamma
\end{cases}
\tag{7.33}
$$

Eq. (7.33) indicates that the critical issue is to calculate the values of $\partial\Phi/\partial d_i$, $\|\nabla\Phi\|$ along the structural boundary $\partial\Omega$ and $\partial\Phi^s/\partial d_i$, $\|\nabla\Phi^s\|$ along the Neumann boundary Γ_N. To proceed, we first consider the derivative of the maximum/minimum function. Suppose X is the maximum function of a set of level-set functions $\{\chi_i\}$, namely, $X = \max(\chi_i)$. The derivative of X can thus be obtained as

$$
\frac{\partial X}{\partial \chi_k} =
\begin{cases}
1 & \max(\chi_i) = \chi_k \\
0 & \text{or else}
\end{cases}
\tag{7.34}
$$

A similar result can be obtained for the minimum function. According to Eq. (7.25), the following relations can be derived.

$$
\begin{cases}
\dfrac{\partial \Phi}{\partial d_i^s} = \dfrac{\partial \Phi}{\partial \overline{\Phi}} \cdot \dfrac{\partial \overline{\Phi}}{\partial d_i^s} + \dfrac{\partial \Phi}{\partial \Phi^s} \cdot \dfrac{\partial \Phi^s}{\partial d_i^s} \\[3mm]
\dfrac{\partial \Phi}{\partial d_i^t} = \dfrac{\partial \Phi}{\partial \Phi^t} \cdot \dfrac{\partial \Phi^t}{\partial d_i^t} \\[3mm]
\|\nabla\Phi\| = \left\| \dfrac{\partial \Phi}{\partial \overline{\Phi}} \nabla\overline{\Phi} + \dfrac{\partial \Phi}{\partial \Phi^s} \nabla\Phi^s + \dfrac{\partial \Phi}{\partial \Phi^t} \nabla\Phi^t \right\|
\end{cases}
\tag{7.35}
$$

where $\partial\Phi/\partial\overline{\Phi}$, $\partial\Phi/\partial\Phi^s$, and $\partial\Phi/\partial\Phi^t$ are determined in accordance with Eq. (7.34).

According to Eq. (7.21), $\partial\Phi^s/\partial d_i^s$ and $\nabla\Phi^s$ can be calculated as

$$
\begin{cases}
\dfrac{\partial \Phi^s}{\partial d_i^s} = \dfrac{\partial \Phi^s}{\partial \varphi} \cdot \dfrac{\partial \varphi}{\partial d_i^s} \\[3mm]
\nabla\Phi^s = \dfrac{\partial \Phi^s}{\partial \varphi} \nabla\varphi + \dfrac{\partial \Phi^s}{\partial \psi} \nabla\psi
\end{cases}
\tag{7.36}
$$

where $\partial\Phi^s/\partial\varphi$ and $\partial\Phi^s/\partial\psi$ are determined by Eq. (7.34). The partial derivatives $\partial\varphi/\partial d_i^s$, $\partial\varphi/\partial x$, and $\partial\varphi/\partial y$ can be calculated as

$$\begin{cases} \dfrac{\partial \varphi}{\partial d_i^s} = \pm B_{i,p}\left(\dfrac{\theta - \theta_L}{\theta_U - \theta_L}\right) \\[4mm] \dfrac{\partial \varphi}{\partial x} = \pm\left[\dfrac{1}{\theta_U - \theta_L}\sum_{j=1}^{n} R_{ij}B_{j,p}^{(1)}\left(\dfrac{\theta - \theta_L}{\theta_U - \theta_L}\right)\dfrac{\partial \theta}{\partial x} - \dfrac{x - x_0}{\sqrt{(x-x_0)^2 + (y-y_0)^2}}\right] \\[4mm] \dfrac{\partial \varphi}{\partial y} = \pm\left[\dfrac{1}{\theta_U - \theta_L}\sum_{j=1}^{n} R_{ij}B_{j,p}^{(1)}\left(\dfrac{\theta - \theta_L}{\theta_U - \theta_L}\right)\dfrac{\partial \theta}{\partial y} - \dfrac{y - y_0}{\sqrt{(x-x_0)^2 + (y-y_0)^2}}\right] \end{cases}$$

$$(7.37)$$

with

$$\frac{\partial \theta}{\partial x} = -\frac{y - y_0}{(x-x_0)^2 + (y-y_0)^2} \quad \frac{\partial \theta}{\partial y} = \frac{x - x_0}{(x-x_0)^2 + (y-y_0)^2} \qquad (7.38)$$

Besides, the following derivative relation holds for $\overline{\Phi}$

$$\frac{\partial \overline{\Phi}}{\partial d_i^s} = -\frac{\partial \Phi^s}{\partial d_i^s}, \frac{\partial \overline{\Phi}}{\partial x} = -\frac{\partial \Phi^s}{\partial x}, \frac{\partial \overline{\Phi}}{\partial y} = -\frac{\partial \Phi^s}{\partial y} \qquad (7.39)$$

In accordance with Eq. (7.23), the derivative of Φ^t with respect to topology design variable d_i^t and coordinate (x, y) can be written as

$$\begin{cases} \dfrac{\partial \Phi^t(x, d^t)}{\partial d_i^t} = -\sum_{i=1}^{n}\left(\dfrac{\partial \Phi^t}{\partial \varphi_i^t} \cdot \dfrac{\partial \varphi_i^t}{\partial d_i^t}\right) = -\dfrac{\partial \Phi^t}{\partial \varphi_i^t} \cdot \dfrac{\partial \varphi_i^t}{\partial d_i^t} \\[4mm] \dfrac{\partial \Phi^t(x, d^t)}{\partial x} = -\sum_{i=1}^{n}\left(\dfrac{\partial \Phi^t}{\partial \varphi_i^t} \cdot \dfrac{\partial \varphi_i^t}{\partial x}\right) \\[4mm] \dfrac{\partial \Phi^t(x, d^t)}{\partial y} = -\sum_{i=1}^{n}\left(\dfrac{\partial \Phi^t}{\partial \varphi_i^t} \cdot \dfrac{\partial \varphi_i^t}{\partial y}\right) \end{cases}$$

$$(7.40)$$

with

$$\begin{cases} \dfrac{\partial \varphi_i^t}{\partial x_i} = \dfrac{1}{2\pi}\sum_{j=1}^{n_t} R_{ij}B_{j,p}^{(1)}\left(\dfrac{\theta + \pi/2}{2\pi}\right)\dfrac{\partial \theta}{\partial x_i} + \dfrac{x - x_i}{\sqrt{(x-x_i)^2 + (y-y_i)^2}} \\[4mm] \dfrac{\partial \varphi_i^t}{\partial y_i} = \dfrac{1}{2\pi}\sum_{j=1}^{n_t} R_{ij}B_{j,p}^{(1)}\left(\dfrac{\theta + \pi/2}{2\pi}\right)\dfrac{\partial \theta}{\partial y_i} + \dfrac{y - y_i}{\sqrt{(x-x_i)^2 + (y-y_i)^2}} \\[4mm] \dfrac{\partial \varphi_i^t}{\partial R_{ij}} = B_{i,p}\left(\dfrac{\theta + \pi/2}{2\pi}\right) \\[4mm] \dfrac{\partial \varphi_i^t}{\partial x} = -\dfrac{\partial \varphi_i^t}{\partial x_i} \\[4mm] \dfrac{\partial \varphi_i^t}{\partial y} = -\dfrac{\partial \varphi_i^t}{\partial y_i} \end{cases}$$

$$(7.41)$$

where

$$\frac{\partial \theta}{\partial x_i} = \frac{y - y_i}{(x - x_i)^2 + (y - y_i)^2} \quad \text{and} \quad \frac{\partial \theta}{\partial y_i} = \frac{x - x_i}{(x - x_i)^2 + (y - y_i)^2}$$

7.2.3 Numerical examples including design-dependent pressure loads

Several examples are studied in this section to illustrate the validity of the presented method. In all examples, suppose the Young's modulus of the solid material is $E = 7 \times 10^{10}$ MPa and the Poisson's ratio is 0.3. The ratio of the Young's modulus of the "ersatz material" to that of the solid material is 1×10^{-3}. The regularized objective function is minimized subjected to the prescribed volume constraint.

Example 7.4 Topology optimization of a structure undergoing snow load
Fig. 7.27 shows a rectangular design domain undergoing a uniform snow load of magnitude 10^4 MPa on the upper surface. The upper pressure boundary is fixed at both ends and initially modeled by an OBS $\varphi(x,y)$ centered at (8mm, 16mm) with 12 control radii, as shown in Fig. 7.28A. In this example, $\psi = 8 - y$ is used for Eq. (7.21). The bandwidth of the nondesignable offset domain is $\Delta = 0.1$ mm. Plane stress assumption is considered and the thickness of the structure is 1mm. The prescribed upper bound of the volume constraint is $\overline{V} = 40$ mm^3.

The rectangular design domain is discretized into a 48×24 finite cell method mesh, as illustrated in Fig. 7.28B. The inner topology boundaries are composed of 14 CBSs with 12 control radii for each of them. There exist 208 design variables, including 12 shape design variables and $14 \times (12 + 2) = 196$ topology design variables while thousands of design variables exist for the discrete level-set method (Xia et al., 2015).

Fixed and moving pressure boundaries are considered as two cases for the sake of comparison. The optimized structure and corresponding finite cell

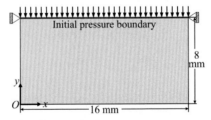

FIG. 7.27 A structure undergoing snow load.

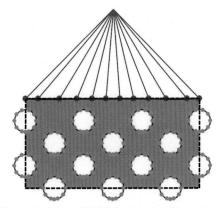

● Control points of OBS for pressure boundary
● Control points of effective CBSs for topology boundary

(A)

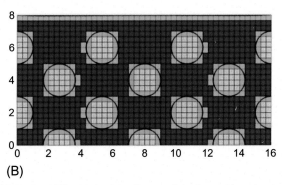

(B)

FIG. 7.28 Initial structure: (A) control points of pressure boundary and inner topology boundaries; and (B) corresponding finite cell model.

method model in the first case are shown in Fig. 7.29 while the solution in the second case is illustrated in Fig. 7.30. The result corresponds to a bowl-like structure with smooth boundaries similar to that obtained in Sigmund and Clausen (2007). The optimized compliances, volumes, and perimeters in both cases are listed in Table 7.4. The optimized compliances in the cases of fixed and moving pressure boundaries are 2.38521 and 2.17403 mJ, respectively. It is seen that the structural compliance is further reduced in the second case owing to the modifiability of the pressure boundary, although no significant benefit is achieved to increase the structural rigidity. Fig. 7.31 shows the convergence histories when the regularization factor is $\beta = 0.01$.

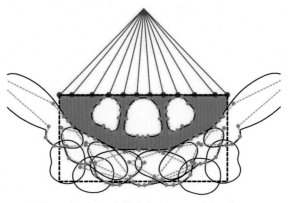

● Control points of OBS for pressure boundary

● Control points of effective CBSs for topology boundary

(A)

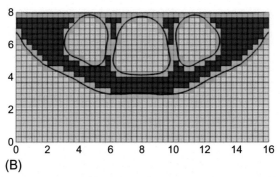

(B)

FIG. 7.29 Optimized structure in the case of fixed pressure boundary: (A) control points of fixed pressure boundary and inner topology boundaries; and (B) corresponding finite cell model.

The problem without perimeter regularization is further solved. Fig. 7.32 shows that three inner holes exist in the optimized result. The compliance is reduced to 2.12618 mJ in comparison with that in Fig. 7.30A. Fig. 7.33 provides intermediate results with and without regularization. Obviously, the perimeter regularization is effective to remove the inner holes during the optimization process.

Example 7.5 Topology optimization of a structure with external air pressure Consider a square domain with a prescribed square hole inside, as shown in Fig. 7.34. The structure is loaded with surrounding air pressure of magnitude 10^4 MPa and simply supported along four segments. Due to the symmetry, only

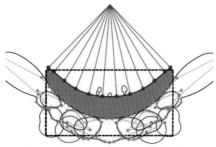

● Control points of OBS for pressure boundary
● Control points of effective CBSs for topology boundary
(A)

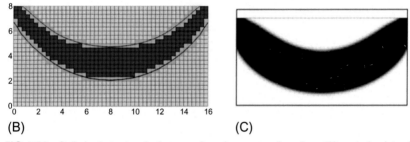

(B)   **(C)**

FIG. 7.30 Optimized structure in the case of moving pressure boundary: (A) control points of moving pressure boundary and inner topology boundaries; (B) corresponding finite cell model; and (C) optimized structure obtained with SIMP model in Sigmund and Clausen (2007).

TABLE 7.4 Comparisons of structural compliance and perimeter for the cases of fixed and moving pressure boundaries.

	Fixed pressure boundary	Moving pressure boundary
Compliance (mJ)	2.38521	2.17403
Volume (mm³)	39.9907	39.9785
Perimeter (mm)	65.1110	37.1481

a quarter of the design domain is optimized with a discretization of 40×40 FCM cells. In this example, plane strain assumption is considered. The volume fraction is limited to 16.667%. The bandwidth of the nondesignable offset domain is $\Delta = 0.5\,\text{mm}$ and the regularization factor $\beta = 0.05$ is used.

Fig. 7.35 shows the initial configuration of the 1/4 structure and corresponding finite cell model. The pressure boundary is modeled with an OBS centered

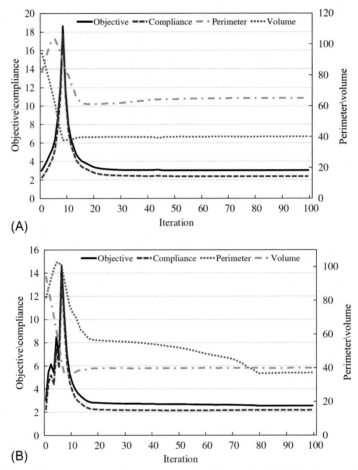

FIG. 7.31 Convergence curves of the structure undergoing snow load: (A) fixed pressure boundary; and (B) moving pressure boundary.

at the origin with 13 control points. The zero-contour of $\psi = \min(40 - x, 40 - y)$ represents the design domain boundary supporting air pressure. The inner topology boundaries are composed of 12 CBSs with 12 control radii for each of them. Thus, a total of 181 design variables exist. After optimization, the 1/4 structure and corresponding finite cell model are shown in Fig. 7.36. The whole structure is shown in Fig. 7.37. It resembles a circular ring and is mechanically reasonable to support external air pressure. The convergence history is illustrated in Fig. 7.38.

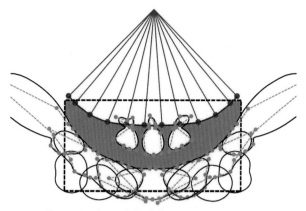

● Control points of OBS for pressure boundary

○ Control points of effective CBSs for topology boundary

FIG. 7.32 Optimized result without perimeter regularization.

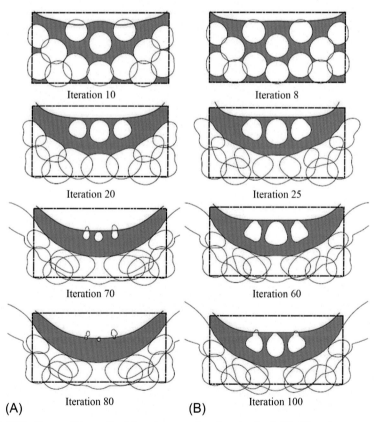

FIG. 7.33 Intermediate results of the structure undergoing snow load: (A) with perimeter regularization; and (B) without perimeter regularization.

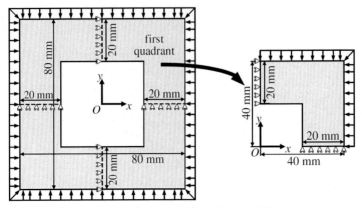

FIG. 7.34 A simply supported structure surrounding with external air pressure.

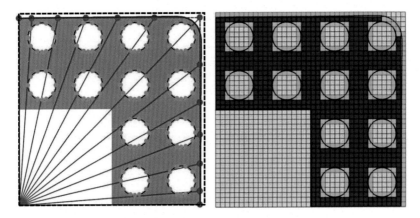

● Control points of OBS for pressure boundary

● Control points of CBSs for topology boundary

(A)   (B)

FIG. 7.35 Initial configuration of 1/4 structure: (A) control points of pressure boundary and inner topology boundaries; and (B) corresponding finite cell model.

Example 7.6 Topology optimization of cyclic symmetry structure subjected to internal pressure

Fig. 7.39 shows the design problem of a cyclic symmetry structure. Assume the inner boundary of the structure is designable and loaded with a uniform pressure of magnitude 10^6 MPa while the outer boundary is fixed. Plane strain assumption is considered. The upper bound of the solid area constraint is 150 mm^2.

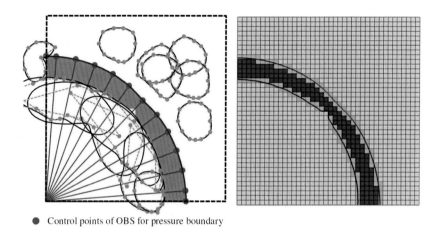

● Control points of OBS for pressure boundary

● Control points of CBSs for topology boundary

(A)   (B)

FIG. 7.36 Optimized results of 1/4 structure: (A) control points of pressure boundary and inner topology boundaries; and (B) corresponding finite cell model.

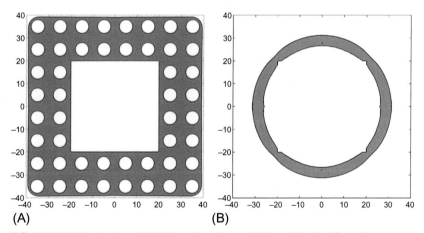

(A)   (B)

FIG. 7.37 Whole structure: (A) initial configuration; and (B) optimized configuration.

The cyclic symmetry structure can be regarded as a periodic array of representative volume elements (RVEs) along the circular direction. The whole structure is embedded in a square design domain that is discretized into a 75×75 finite cell grid. The bandwidth of the nondesignable offset domain is $\Delta = 0.5\,\mathrm{mm}$ and the regularization factor is $\beta = 0.1$.

First, suppose the whole structure is composed of eight identical representative volume elements. The moving pressure boundary in each representative

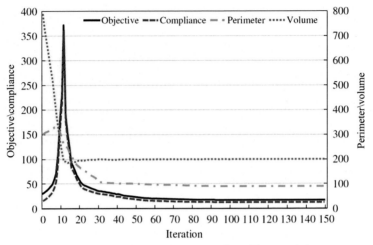

FIG. 7.38 Convergence history of the structure surrounding with external air pressure.

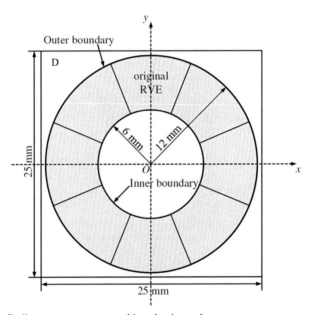

FIG. 7.39 Cyclic symmetry structure subjected to internal pressure.

volume element is modeled with an OBS centered at the origin and including six control radii. The variation range of each control radius is (6.5 mm, 9 mm). In addition, the inner topology of each representative volume element is described by a set of CBSs with 12 control radii for each of them. Three initial structural configurations are considered to study their effects. In the first two cases,

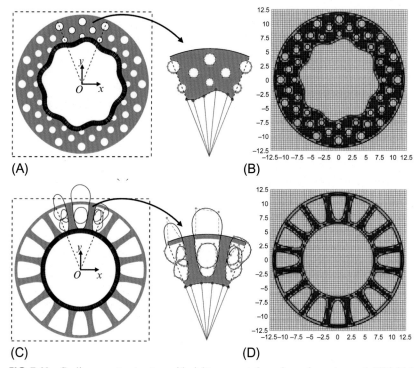

FIG. 7.40   Cyclic symmetry structure with eight representative volume elements-case 1: (A) initial structure and representative volume element; (B) initial finite cell model; (C) optimized structure and representative volume element; and (D) optimized finite cell model.

suppose the inner topologies are the same and described by eight CBSs while the initial pressure boundary is modeled by an OBS with different control radii $\{R_j\} = [7,7.25,7.5,6.75,6.5,7]$ and $\{R_j\} = [6.5,6.7,7.1,7.1,6.7,6.5]$, respectively. Figs. 7.40 and 7.41 show the initial and optimized structures and corresponding finite cell models. It is seen that the control radii $\{R_j\}$ of the pressure boundaries reach the lower bound value of 6.5 mm in both cases. This is reasonable because the smaller the control radii are, the less the total pressure loads will be applied. The optimized topologies are similar and composed of 16 radial members to bear the radial pressure.

In the third case, suppose the initial pressure boundary is the same as in the first case while the inner topology is composed of 10 CBSs, as shown in Fig. 7.42. Likewise, the control radii of the pressure boundary reach the lower bound value for the optimized structure and 24 perpendicular members are generated. The optimized compliances, volumes, and perimeters in three cases are listed in Table 7.5. The optimized compliances are 185.232, 185.204, and 175.826 mJ, respectively. It is indicated that a finer configuration results in a

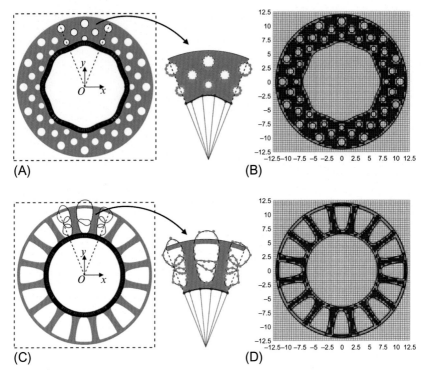

FIG. 7.41 Cyclic symmetry structure with eight representative volume elements-case 2: (A) initial structure and representative volume element; (B) initial finite cell model; (C) optimized structure and representative volume element; and (D) optimized finite cell model.

better rigidity for the optimized structure. This implies that the optimized results are relevant to the initial configuration and are not rigorously cyclic symmetry due to the unsymmetry of the initial configurations. Fig. 7.43 shows the convergence histories in all three cases.

Finally, the test is made for only one representative volume element. The pressure boundary in the representative volume element is described by a CBS with 46 control radii. The inner topology is defined by 32 CBSs with 12 control radii for each of them. The configurations and finite cell models for both the initial and optimized structures are shown in Fig. 7.44. The control radii of the pressure boundary reach 6.5 mm and the inner topology consists of 14 radial members. The optimized compliance is 189.264 mJ. Convergence curves are shown in Fig. 7.45.

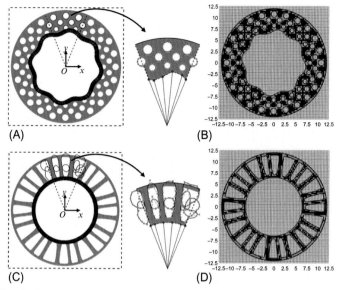

FIG. 7.42 Cyclic symmetry structure with eight representative volume elements-case 3: (A) initial structure and representative volume element; (B) initial finite cell model; (C) optimized structure and representative volume element; and (D) optimized finite cell model.

TABLE 7.5 Comparisons of structural compliances, volumes, and perimeters for different cases.

	Case 1 (8 × 8 CBSs)	Case 2 (8 × 8 CBSs)	Case 3 (8 × 10 CBSs)
Compliance (mJ)	185.232	185.204	175.826
Volume (mm³)	149.970	150.015	149.984
Perimeter (mm)	319.984	320.697	390.031

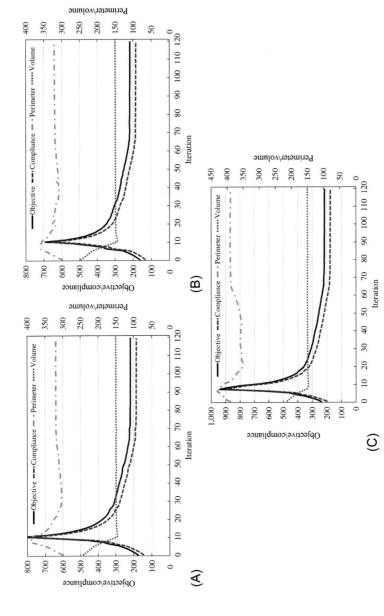

FIG. 7.43 Convergence curves of the cyclic symmetry structure with eight representative volume elements: (A) case 1; (B) case 2; and (C) case 3.

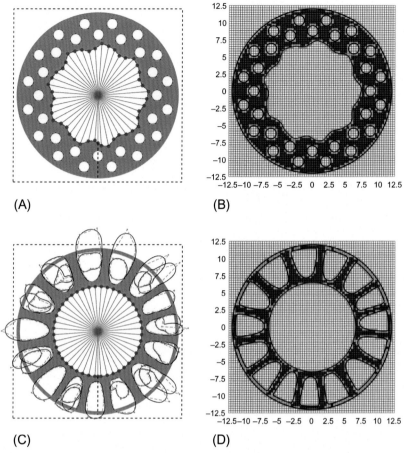

FIG. 7.44 Cyclic symmetry structure with one representative volume element: (A) initial structure and representative volume element; (B) initial finite cell model; (C) optimized structure and representative volume element; and (D) optimized finite cell model.

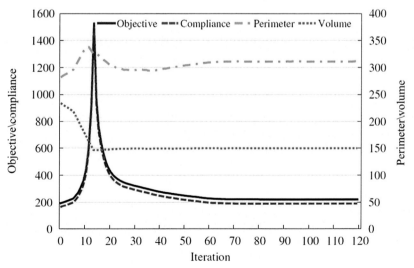

FIG. 7.45 Convergence curves of the cyclic symmetry structure with one representative volume element.

References

Allaire, G., Jouve, F., Toader, A.M., 2004. Structural optimization using sensitivity analysis and a level-set method. J. Comput. Phys. 194 (1), 363–393.

Ansola, R., Canales, J., Tárrago, J.A., 2006. An efficient sensitivity computation strategy for the evolutionary structural optimization (ESO) of continuum structures subjected to self-weight loads. Finite Elem. Anal. Des. 42 (14–15), 1220–1230.

Bendsøe, M.P., Kikuchi, N., 1988. Generating optimal topologies in structural design using a homogenization method. Comput. Methods Appl. Mech. Eng. 71 (2), 197–224.

Bendsøe, M.P., Sigmund, O., 1999. Material interpolation schemes in topology optimization. Arch. Appl. Mech. 69 (9–10), 635–654.

Bendsøe, M.P., Sigmund, O., 2013. Topology optimization: Theory, Methods, and Applications, second ed. Springer, Berlin.

Bourdin, B., Chambolle, A., 2003. Design-dependent loads in topology optimization. ESAIM Control Optim. Calc. Var. 9, 19–48.

Bruggi, M., Cinquini, C., 2009. An alternative truly-mixed formulation to solve pressure load problems in topology optimization. Comput. Methods Appl. Mech. Eng. 198 (17–20), 1500–1512.

Bruyneel, M., Duysinx, P., 2005. Note on topology optimization of continuum structures including self-weight. Struct. Multidiscip. Optim. 29 (4), 245–256.

Chen, B.C., Kikuchi, N., 2001. Topology optimization with design-dependent loads. Finite Elem. Anal. Des. 37 (1), 57–70.

Du, J., Olhoff, N., 2004a. Topological optimization of continuum structures with design-dependent surface loading—part I: new computational approach for 2D problems. Struct. Multidiscip. Optim. 27 (3), 151–165.

Du, J., Olhoff, N., 2004b. Topological optimization of continuum structures with design-dependent surface loading—part II: algorithm and examples for 3D problems. Struct. Multidiscip. Optim. 27 (3), 166–177.

Eschenauer, H.A., Kobelev, V.V., Schumacher, A., 1994. Bubble method for topology and shape optimization of structures. Struct. Optim. 8 (1), 42–51.

Fuchs, M.B., Shemesh, N.N.Y., 2004. Density-based topological design of structures subjected to water pressure using a parametric loading surface. Struct. Multidiscip. Optim. 28 (1), 11–19.

Gao, T., Zhang, W., 2010. Topology optimization involving thermo-elastic stress loads. Struct. Multidiscip. Optim. 42 (5), 725–738.

Gao, T., Xu, P., Zhang, W., 2016. Topology optimization of thermo-elastic structures with multiple materials under mass constraint. Comput. Struct. 173, 150–160.

Guo, X., Zhao, K., Gu, Y., 2004. Topology optimization with design-dependent loads by level set approach. In: 10th AIAA/ISSMO Multidisciplinary Analysis and Optimization Conference. Albany, New York, pp. 1–10.

Hammer, V.B., Olhoff, N., 2000. Topology optimization of continuum structures subjected to pressure loading. Struct. Multidiscip. Optim. 19 (2), 85–92.

Huang, X., Xie, Y.M., 2011. Evolutionary topology optimization of continuum structures including design-dependent self-weight loads. Finite Elem. Anal. Des. 47 (8), 942–948.

Kumar, A.V., Parthasarathy, A., 2011. Topology optimization using B-spline finite elements. Struct. Multidiscip. Optim. 44, 471–481.

Lee, E., Martins, J.R., 2012. Structural topology optimization with design-dependent pressure loads. Comput. Methods Appl. Mech. Eng. 233, 40–48.

Parvizian, J., Düster, A., Rank, E., 2007. Finite cell method. Comput. Mech. 41 (1), 121–133.

Qian, X., 2013. Topology optimization in B-spline space. Comput. Methods Appl. Mech. Eng. 265, 15–35.

Radovcic, Y., Remouchamps, A., 2002. BOSS QUATTRO: an open system for parametric design. Struct. Multidiscip. Optim. 23 (2), 140–152.

Rodrigues, H., Fernandes, P., 1995. A material based model for topology optimization of thermo-elastic structures. Int. J. Numer. Methods Eng. 38 (12), 1951–1965.

Seo, Y.D., Kim, H.J., Youn, S.K., 2010. Isogeometric topology optimization using trimmed spline surfaces. Comput. Methods Appl. Mech. Eng. 199 (49–52), 3270–3296.

Sigmund, O., Clausen, P.M., 2007. Topology optimization using a mixed formulation: an alternative way to solve pressure load problems. Comput. Methods Appl. Mech. Eng. 196 (13–16), 1874–1889.

Sigmund, O., Maute, K., 2013. Topology optimization approaches. Struct. Multidiscip. Optim. 48 (6), 1031–1055.

Stolpe, M., Svanberg, K., 2001. An alternative interpolation scheme for minimum compliance topology optimization. Struct. Multidiscip. Optim. 22 (2), 116–124.

Svanberg, K., 1995, May. A globally convergent version of MMA without linesearch. In: Proceedings of the First World Congress of Structural and Multidisciplinary Optimization, Goslar, Germany, vol. 28, pp. 9–16.

Turteltaub, S., Washabaugh, P., 1999. Optimal distribution of material properties for an elastic continuum with structure-dependent body force. Int. J. Solids Struct. 36 (30), 4587–4608.

Wang, M.Y., Wang, X., Guo, D., 2003. A level set method for structural topology optimization. Comput. Methods Appl. Mech. Eng. 192 (1–2), 227–246.

Wang, C., Zhao, M., Ge, T., 2016. Structural topology optimization with design-dependent pressure loads. Struct. Multidiscip. Optim. 53 (5), 1005–1018.

Wei, P., Wang, M.Y., 2009. Piecewise constant level set method for structural topology optimization. Int. J. Numer. Methods Eng. 78 (4), 379–402.

Xavier, M., Novotny, A.A., 2017. Topological derivative-based topology optimization of structures subject to design-dependent hydrostatic pressure loading. Struct. Multidiscip. Optim. 56 (1), 47–57.

Xia, Q., Shi, T., 2016. Optimization of structures with thin-layer functional device on its surface through a level set based multiple-type boundary method. Comput. Methods Appl. Mech. Eng. 311, 56–70.

Xia, Q., Wang, M.Y., 2008. Topology optimization of thermoelastic structures using level set method. Comput. Mech. 42, 837–857.

Xia, Q., Wang, M.Y., Shi, T., 2015. Topology optimization with pressure load through a level set method. Comput. Methods Appl. Mech. Eng. 283, 177–195.

Xu, H., Guan, L., Chen, X., Wang, L., 2013. Guide-weight method for topology optimization of continuum structures including body forces. Finite Elem. Anal. Des. 75, 38–49.

Yamada, T., Izui, K., Nishiwaki, S., Takezawa, A., 2010. A topology optimization method based on the level set method incorporating a fictitious interface energy. Comput. Methods Appl. Mech. Eng. 199 (45–48), 2876–2891.

Yamada, T., Izui, K., Nishiwaki, S., 2011. A level set-based topology optimization method for maximizing thermal diffusivity in problems including design-dependent effects. J. Mech. Des. 133 (3), 031011.

Yang, X.Y., Xie, Y.M., Steven, G.P., 2005. Evolutionary methods for topology optimisation of continuous structures with design dependent loads. Comput. Struct. 83 (12–13), 956–963.

Zhang, H., Zhang, X., Liu, S., 2008. A new boundary search scheme for topology optimization of continuum structures with design-dependent loads. Struct. Multidiscip. Optim. 37 (2), 121–129.

Zheng, B., Chang, C.J., Gea, H.C., 2009. Topology optimization with design-dependent pressure loading. Struct. Multidscip. Optim. 38, 535–545.

Zhang, W., Zhu, J., Gao, T., 2016. Topology Optimization in Engineering Structure Design. ISTE Press Ltd, Elsevier, London.

Zhu, J., Zhang, W., Beckers, P., 2009. Integrated layout design of multi-component system. Int. J. Numer. Methods Eng. 78 (6), 631–651.

Zhu, J.H., Zhang, W.H., Xia, L., 2016. Topology optimization in aircraft and aerospace structures design. Arch. Comput. Methods Eng. 23 (4), 595–622.

Chapter 8

Integration of feature-driven optimization with additive manufacturing

Additive manufacturing (AM) is an emerging technique that provides great flexibility for the fabrication of freeform structures with complicated configurations. In recent years, AM receives significant attention (Frazier, 2014; Horn and Harrysson, 2012; Wong and Hernandez, 2012). Compared to the conventional techniques of subtractive manufacturing, an AM process such as selective laser melting (SLM) is characteristic of depositing materials from bottom to top layer by layer. Therefore, AM is considered a better choice to achieve seamless integration with the advanced topology optimization method to attain lightweightness and high performance of a structure, both in design and manufacturing.

Although AM significantly opens up the design space, it is not completely a freeform manufacturing technique. For example, enclosed voids in a structure should be avoided to evacuate the rest of the unmelted powders in powder-based processes such as SLM (Li et al., 2016). Building accuracy, interlayer mechanical properties, and surface finish are also important factors influencing the manufacturing quality of a structure (Diegel et al., 2010; Gao et al., 2015; Qattawi and Ablat, 2017). Especially, additional supports should be taken into account at the design stage to prevent overhang portions of a structure from collapsing in the AM process. Additional supports also have the functionality of dissipating heat into the building platform to avoid local distortion of the structure. Nevertheless, as these supports are finally removed, the waste of materials and time cost are evident and even the structural surface quality would be deteriorated if postprocessing operations were not carefully made.

Within the framework of feature-driven optimization (FDO), this chapter is focused upon two major concerns related to AM, that is, designs of self-supporting structures without the need for additional support and structures without enclosed voids.

The Feature-driven Method for Structural Optimization. https://doi.org/10.1016/B978-0-12-821330-8.00008-0

8.1 Topology optimization of self-supporting structure with polygon features

Much effort has been devoted to reducing additional supports in the AM process. One simple solution is to optimize the build direction. This is figured out by comparing the support volumes among all possible orientations (Strano et al., 2013; Calignano, 2014; Das et al., 2015; Morgan et al., 2016). Although this kind of method is able to reduce supports to some extent, it is limited by the geometry of the predesigned part itself. The second kind of method is to exploit the topologies or materials of supports to reduce the weight while the functionality and printability are ensured (Hussein et al., 2013; Calignano, 2014; Dumas et al., 2014; Vanek et al., 2014; Barnett and Gosselin, 2015). This kind of method is still unable to completely eliminate additional supports. An alternative is to design self-supporting structures. Hu et al. (2015) proposed an orientation-driven shape optimization method to alter models. Specifically, after a model is divided into tetrahedral meshes, reorientation is taken on the tetrahedra with overhang surface facets while ensuring the minimal shape variation. Leary et al. (2014) improved initially optimized results with additional structures as part of the design that will not be removed. However, both methods are postprocessing approaches that will alter the weight and perturb the optimality of the original designs.

Recently, more and more researchers have shown interest in integrating the AM process with topology optimization (Brackett et al., 2011; Gaynor et al., 2014; Clausen, 2016; Gaynor and Guest, 2016; Langelaar, 2016, 2017; Mirzendehdel and Suresh, 2016; Zegard and Paulino, 2016; Guo et al., 2017). To achieve a self-supporting structure, Brackett et al. (2011) proposed a penalty function to quantify the overall violation of self-supporting requirements in the framework of bidirectional evolutionary structural optimization. Gaynor and Guest (2016) adopted a series of projection operations to enforce the requirement of minimum length scale and satisfy overhang constraints. In the AM process, as the presence of materials in a given layer depends upon the material distribution in the layer(s) below, topology determination and sensitivity analysis must proceed in a layer-by-layer manner, which is generally inefficient from a computational point of view. Similarly, Langelaar (2016, 2017) implemented a layerwise nonlinear spatial filter in the framework of density-based topology optimization. Materials are assigned to an element, provided it is sufficiently supported by the elements in the lower layer. This method indeed produces self-supporting structures, but is limited to the rectangular mesh and the specific critical overhang angle $\beta_0 = 45°$.

The feature-driven optimization (FDO) method presented in Chapter 6 is an alternative promising approach. In this section, a kind of polygon-featured hole is introduced as a design primitive to drive topology variation. To control the overhang angles, ratio design variables are formulated. V-shaped areas are checked automatically in the optimized result for the presence. Intersecting polygons are correspondingly modified, replaced with new polygons, and further optimized until all V-shaped areas are eliminated.

8.1.1   Representation of polygon-featured holes

Take the model shown in Fig. 8.1A as an example. Assume **b** represents the build direction. The green solid and the red dashed boundaries of the circular hole represent the overhang portions. A critical issue for a structure to be self-supported is that the inclined angles along the overhang boundaries are greater than or at least equal to the critical overhang angle β_0 (Gaynor and Guest, 2016). Notice that β_0 is an acute angle whose value depends upon the specific AM process (Mertens et al., 2014; Kranz et al., 2015; Zegard and Paulino, 2016). During the AM process, the red arc area may easily collapse without supports because all the angles defined by the corresponding tangents are smaller than β_0. On the contrary, the structure shown in Fig. 8.1B satisfies the requirement of the critical overhang angle everywhere and no supports are required.

A polygon is composed of a finite number of straight line segments. The inclined angle of each segment is easily computed to facilitate the control of overhang angles for the structure to be self-supported. In this section, some rules are imposed for the utilization of polygon features. To simplify the discussion, Fig. 8.2 shows an 11-side polygon and a 12-side polygon to represent polygons of odd and even numbers of sides, respectively. OV_1 (O is the center point and V_i is the ith vertex of a polygon) should have the same orientation as the build direction **b**. The subscript of V_i is incremented counter-clockwise and the angle α between any two adjacent vectors is constant with $\alpha = 32.73°$ and $\alpha = 30°$, respectively.

In order to maintain the overhang angle condition $\beta_i \geq \beta_0$ for the ith side, the lengths of OV_i and OV_{i+1} can be adjusted in design. When the ratio of the two lengths is controlled within a certain range, this condition will be fulfilled. To do this, the critical ratio $\overline{\lambda}_i$ related to $\beta_i = \beta_0$ is calculated according to three interior angles of triangle OV_iV_{i+1} and the Law of Sines. In this section, the length ratio λ_i between OV_i and OV_{i+1} is formulated as a design variable with upper bound $\overline{\lambda}_i$. In this way, the overhang angle condition can be taken into account directly and concisely.

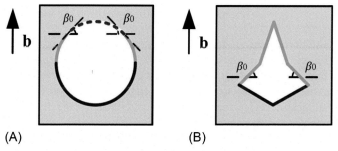

FIG. 8.1 Definition of the critical overhang angle: (A) the overhang portions in a structure require supports; and (B) the overhang portions in a structure do not require supports.

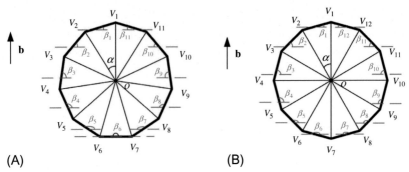

FIG. 8.2 Definition of polygons with odd and even sides: (A) an 11-side polygon; and (B) a 12-side polygon.

However, there could exist a case where the critical ratio does not exist. For instance, when $\beta_0 = 45°$, the inclined angle β_6 of V_6V_7 in Fig. 8.2A is always obtuse and cannot be equal to β_0. This means that no matter how the lengths of OV_6 and OV_7 change, V_6V_7 will never be located in overhang portions and never violate the overhang angle condition. Therefore, it is unnecessary to formulate and bound the related ratio design variable λ_6. Instead, the two lengths of OV_6 and OV_7 are directly considered as design variables. In addition, design variables also include the coordinates of the center point. The design variables involved in the 11-side polygon and 12-side polygon are $\{l_1, \lambda_1, \lambda_2, \lambda_3, \lambda_4, l_6, l_7, \lambda_8, \lambda_9, \lambda_{10}, \lambda_{11}, x_0, y_0\}$ and $\{l_1, \lambda_1, \lambda_2, \lambda_3, \lambda_4, l_6, l_7, l_8, \lambda_9, \lambda_{10}, \lambda_{11}, \lambda_{12}, x_0, y_0\}$, respectively. Among them, l_i is the length of OV_i, x_0 and y_0 are the coordinates of the center point, and λ_i is defined as follows

$$\lambda_i = \begin{cases} l_{i+1}/l_i & \text{if } i < n/2 \\ l_i/l_{i+1} & \text{if } i > n/2 \text{ and } i \neq n \\ l_i/l_1 & \text{if } i = n \end{cases} \tag{8.1}$$

where n is the total number of sides of a polygon.

This relation is valid for an arbitrary n-side polygon defined above. In the optimization process, each design variable can be changed independently. It is observed that apart from the number of sides of a polygon, the value of β_0 also influences the critical ratios and may further change the numbers of ratio design variables and length design variables. So, once β_0 changes, the design variables of a polygon need to be redefined.

For comparison, freeform topology optimization is also studied. In this case, design variables involved in an n-side polygon are defined as $\{l_1, l_2, ..., l_{n-1}, l_n, x_0, y_0\}$ because there is no longer a requirement of the critical overhang angle.

8.1.2 Construction of the level-set function for a polygon-featured hole with Boolean operations

In this section, the level-set function of a polygon feature is also realized by Boolean operations. Two methods are discussed below.

● **Side-based method**

When all sides are viewed as basic elements of a polygon, the level-set function Φ_p of an n-side polygon can be expressed as follows

$$\Phi_p = \min\left(\Phi_{s1}, \Phi_{s2}, \ldots, \Phi_{sn}\right) \tag{8.2}$$

where Φ_{si} is the level-set function of the ith side and it can be obtained according to the coordinates (x_i, y_i) and (x_{i+1}, y_{i+1}) of the two endpoints V_i and V_{i+1}. To ensure the closure of the polygon, the value of Φ_{si} at the center point (x_0, y_0) should be greater than zero. The general form is

$$\Phi_{si} = \begin{cases} ax + by + c & \text{if } ax_0 + by_0 + c > 0 \\ -(ax + by + c) & \text{if } ax_0 + by_0 + c < 0 \end{cases} \tag{8.3}$$

with

$$a = y_{i+1} - y_i, \quad b = x_i - x_{i+1}, \quad c = y_i x_{i+1} - x_i y_{i+1} \tag{8.4}$$

However, this method is only suitable to convex polygon features. For a concave polygon feature, the resulting polygon is not the desired one, as shown in Fig. 8.3. The black solid lines define a 12-side polygon while the blue shaded area is the constructed polygon feature.

● **Triangle-based method**

An n-side polygon feature can be regarded as the union of n triangles. Each triangle can be constructed by the side-based method presented above. Then, the level-set function Φ_p of a polygon with n sides can be expressed as:

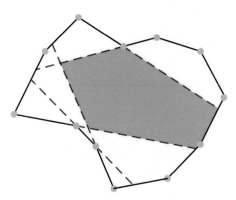

FIG. 8.3 The geometry constructed by the side-based method.

$$\Phi_p = \max\left(\Phi_{t1}, \Phi_{t2}, \dots, \Phi_{tn}\right) \tag{8.5}$$

where Φ_{ti} is the level-set function of the ith triangle, composed of the level-set functions of OV_i, OV_{i+1}, and V_iV_{i+1}. Obviously, this triangle-based method is also suitable to the description of concave polygon features. Fig. 8.4 shows that a polygon bounded by black lines is identical to the constructed geometry in shade. Theoretically, although triangle $OV_{i-1}V_i$ and triangle OV_iV_{i+1} share the common overlapping edge OV_i, such an overlap can be eliminated by utilizing the KS function.

For a rectangular design domain involving m polygon features, the level-set functions Φ of the whole structure can be expressed as

$$\Phi = \min\left(\Phi_r, -\max\left(\Phi_{P1}, \Phi_{P2}, \dots, \Phi_{Pm}\right)\right) \tag{8.6}$$

where $\Phi_r(x)$ is the level-set function of the rectangular design domain. $\Phi_{P1}(x)$, $\Phi_{P2}(x),\dots,\Phi_{Pm}(x)$ denote the level-set functions of m polygon features inside the design domain. The negative sign means that each polygon represents a hole rather than a solid inclusion.

The maximization and minimization can approximately be represented by an envelope function such as the KS function, the R function, and P-norm. In the following sections, the KS function is utilized to construct polygon features and structures because of its simple operation.

8.1.3   Elimination of unprintable V-shaped areas caused by intersecting polygon-featured holes

From the above presentation, we can see that with ratio design variables, the inclined angles of all overhang portions can be effectively controlled by β_0. However, the optimized result cannot avoid the unprintable case of V-shaped areas. As

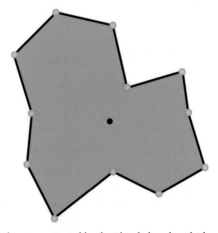

FIG. 8.4   A 12-side polygon constructed by the triangle-based method.

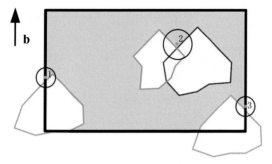

FIG. 8.5   V-shaped areas in a part.

shown in Fig. 8.5, although the optimized structure satisfies the overhang angle condition, the V-shaped areas do exist and need supports from below to realize the structure by the AM process. The presence of V-shaped areas is due to the limitation of the inclined angles at the intersection points when two polygons intersect or polygons intersect with the design domain.

To solve this problem, V-shaped areas are first identified and locally modified. Reoptimization is further carried out. The detailed steps are as follows.

- **Step 1: Identifying V-shaped areas.** It is very easy to determine the intersecting polygons within a design domain, when vertex coordinates of two polygons are provided. In the same way, intersections between a polygon hole and the left or right boundary of the design domain are obtained. Because V-shaped areas could be formed only at the intersections located at the overhang portions, a test point is introduced at each intersection point along the build direction with a small offset distance, just like the three blue dots in Fig. 8.5. Then, the value of the level-set function Φ related to the whole structure is evaluated at each test point. $\Phi \geq 0$ means the corresponding intersection point is located at the overhang portions. To make things clear, take the structure in Fig. 8.6 as an example. Red triangles denote

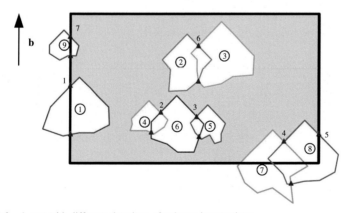

FIG. 8.6   A part with different situations of polygon intersections.

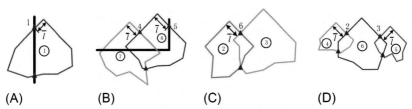

FIG. 8.7 The definition of $\bar{l}$ at intersections in Fig. 8.6: (A) point 1; (B) point 4 and point 5; (C) point 6; and (D) point 2 and point 3.

the intersections between two polygons or between a polygon and the left or right boundary of the design domain. After defining the test points above these intersections and checking the corresponding level-set function values, the intersection points 1–7 are preserved with $\Phi \geq 0$. Because the center of polygon 9 is outside the domain, point 7 is not a point for the production of a V-shaped area and is further eliminated.

So far, all the doubtful intersection points where V-shapes may appear have been recognized. Now, a length $\bar{l}$ will be evaluated at each of these intersection points to judge whether there exists a V-shaped area. $\bar{l} = 0$ means that no V-shaped area occurs at the intersection point. Fig. 8.7 illustrates the definition of $\bar{l}$ at points 1–6. For points 1 and 5 located on the contour of the design domain, $\bar{l}$ represents the distance from point 1 to the first vertex V_1 of polygon 1 and the distance from point 5 to the first vertex V_1 of polygon 8, respectively. The remaining questionable points 2, 3, 4, 6 are all the intersections of two polygons. The distances from the intersection point to V_1 of the two intersecting polygons are separately calculated and the shorter one is selected as $\bar{l}$. If the sum of lengths $\bar{l}$ at all doubtful points is very small, V-shaped areas can be ignored and the optimized result is recognized as the final solution. Otherwise, we have to go to the next step.

- **Step 2: Polygon modification and reoptimization.** The identified intersections related to the V-shaped areas will be handled as follows. When a V-shaped area is formed by the intersection of two or more polygons, the intersecting polygons will be replaced by a new polygon with a similar shape. Take the two intersecting polygons in Fig. 8.8 for instance. The

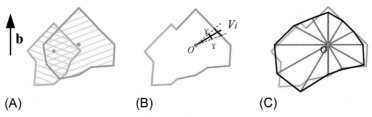

FIG. 8.8 The replacement of two intersecting polygons with a new polygon: (A) two intersecting polygons; (B) determination of vertex positions of the new polygon; and (C) the new polygon.

center of the blue polygon in Fig. 8.8A will be selected as the center of the new polygon because the first vertex V_1 of the blue polygon is higher than that of the green polygon along the build direction. Moreover, in order to avoid the redefinition of design variables, the new polygon keeps the same number of sides as the two original polygons. Then, the directions of all OV_i of the new polygon are further determined and the length OV_i is defined in Fig. 8.8B as the average distance from some discrete points on the red line to the center point. The red line is part of the hole boundary between the two gray dotted lines whose directions are related to OV_i and γ. In this process, the function CONTOUR of MATLAB is used to get the coordinates of these discrete points. Fig. 8.8C shows the intersecting polygons replaced with a new one. The method is also applicable to three or more intersecting polygons.

In particular, if several polygons intersect and one of them also intersects with the left or right contour of the design domain, such as polygon 7 and 8 in Fig. 8.6, the new polygon is formed with the same method and then translated perpendicularly to the build direction until the center point is located on the right boundary of the design domain. This process is illustrated in Fig. 8.9. Moreover, in order to avoid the recurrence of the V-shaped area, the center point of the new polygon will not be allowed into the design domain in the next round of optimization.

For the case of polygon 1 intersecting with the domain contour in Fig. 8.6, the polygon is only translated to relocate the center point on the left side of the rectangular domain. Similarly, its center point will be limited outside the design domain in the new round of optimization. Finally, all the polygons associated with the questionable intersections are modified and the V-shaped areas are eliminated. As the number of design variables experiences a decrease due to the reduction of the number of polygons in the structure, the computing burden is lightened in the following reoptimization. Reoptimization will continue until no V-shaped area can be detected within optimized results. Fig. 8.10 illustrates the whole process for the topology optimization of self-supporting structures.

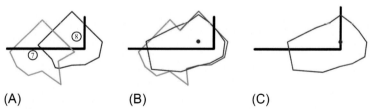

(A)   (B)   (C)

FIG. 8.9 The treatment to polygon 7 and polygon 8 in Fig. 8.6: (A) polygon 7 and polygon 8 in Fig. 8.6; (B) the replacement of two intersecting polygons with a new polygon; and (C) the translation of the new polygon.

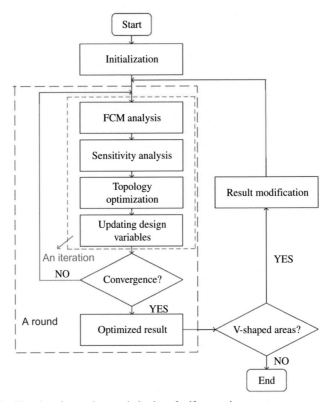

FIG. 8.10 Flowchart for topology optimization of self-supporting structures.

8.1.4 Numerical examples

In this section, the proposed topology optimization method is validated by using several numerical examples. The geometric data and loads are all dimensionless in the following examples. The plane stress state (with unit thickness) is assumed. The Young's modulus and Possion's ratio are $E=1$ and $\nu=0.3$, respectively. In addition, a volume fraction of 50% is used as the upper bound of the volume constraint in the minimization of structural compliance.

Example 8.1 A short beam with a vertical load at the bottom corner
As shown in Fig. 8.11, the short beam is completely fixed along the left side and a vertical force is applied at the bottom corner of the right side. The beam length is $L=12$ and the height is half the length. The structure is discretized into a 120×60 FCM mesh.

The polygon features are used as designable holes to realize the topological changes of the design domain. In the following, we let $\beta_0 = 45°$. Six cases are

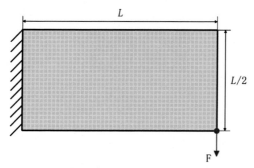

FIG. 8.11 A short beam with a vertical load at the bottom corner.

considered in Table 8.1 and initial layouts of polygon holes are shown in Fig. 8.12 to study the influences upon the optimized topology. The initial structures will be used to perform both freeform topology optimization and topology optimization of self-supporting structures. Fig. 8.13 illustrates the vertex distributions of the 24-side and the 40-side polygons.

For freeform topology optimization, the lengths of all OV_i of each polygon and the coordinates of the center point are considered as design variables to control its position, shape, and size. Fig. 8.14 gives the results of the freeform optimization in six different cases. It is observed that more design variables contribute to the appearance of refined structures, which results in a decrease in compliance. In addition, Fig. 8.14C and F illustrates that in addition to some polygons working on the structural topology, there are useless polygons running

TABLE 8.1 Six design cases for the short beam.

Items	Number of polygons	Number of sides for each polygon	Number of design variables for each polygon	Total number of design variables
Case 1	6	24	26	156
Case 2	10	24	26	260
Case 3	17	24	26	442
Case 4	6	40	42	252
Case 5	10	40	42	420
Case 6	17	40	42	714

FIG. 8.12 The initial layouts of polygons within the design domain: (A) case 1 and case 4 ($J=7599.55$, $A=59.97$); (B) case 2 and case 5 ($J=7800.50$, $A=53.96$); and (C) case 3 and case 6 ($J=19,601.70$, $A=48.44$).

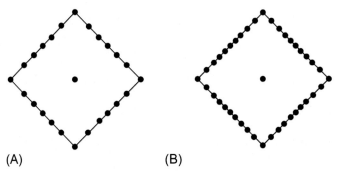

FIG. 8.13 The vertex distribution of a polygon in the initial structure: (A) a 24-side polygon; and (B) a 40-side polygon.

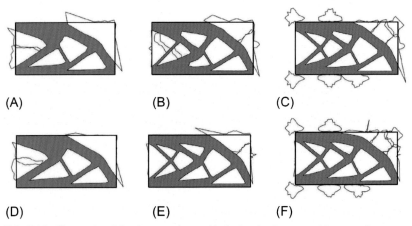

FIG. 8.14 The results of freeform topology optimization in six cases without overhang constraints: (A) case 1 ($J=7170.72$, $A=36.00$); (B) case 2 ($J=7132.11$, $A=36.00$); (C) case 3 ($J=7080.63$, $A=36.00$); (D) case 4 ($J=7151.02$, $A=36.00$); (E) case 5 ($J=7091.44$, $A=36.00$); and (F) case 6 ($J=7062.40$, $A=36.00$).

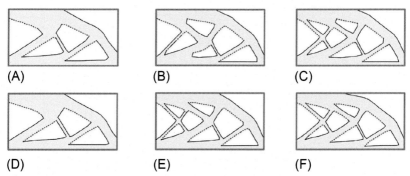

(A)   (B)   (C)

(D)   (E)   (F)

FIG. 8.15 The identified boundaries requiring additional supports within the freeform optimized structures: (A) case 1; (B) case 2; (C) case 3; (D) case 4; (E) case 5; and (F) case 6.

out of the design domain. Fig. 8.16 gives the convergence histories of compliance and volume, correspondingly. At the beginning of the iterations, the compliance changes sharply due to the violation of volume constraint.

However, there exist many violations of overhang constraints in the results of freeform optimization and these portions are highlighted with red dashed lines in Fig. 8.15. During the AM process, supports are required at these places to reinforce the structures.

For the topology optimization of self-supporting structures, the design variables are based on the critical ratios listed in Table 8.2. The 24-side polygon has 16 ratio design variables $\{\lambda_1, \lambda_2, ..., \lambda_8, \lambda_{17}, ..., \lambda_{18}, \lambda_{24}\}$, eight length design variables $\{l_1, l_{10}, l_{11}, ..., l_{16}\}$, and design variables related to the coordinates of the center point $\{x_0, y_0\}$. Similarly, design variables for the 40-side polygon are $\{x_0, y_0, l_1, \lambda_1, \lambda_2, ..., \lambda_{14}, l_{16}, l_{17}, ..., l_{26}, \lambda_{27}, \lambda_{28}, ..., \lambda_{40}\}$. Fig. 8.17 shows the results of the first round topology optimization in six cases. It can be observed that with the same initial number of polygon features, the first round optimized results are of identical topologies. Compared with freeform topology optimization, the overhang angles are significantly controlled in Fig. 8.17. The values of structural compliance correspond to 7586.25, 7377.07, 7343.95, 7523.41, 7362.50, and 7331.19.

However, these results have to be modified and further reoptimized due to the presence of V-shaped areas. Intersecting polygons are correspondingly replaced with a polygon of similar shape and the polygons forming the V-shapes at the boundaries of the design domain are translated. In this process, the number of polygons will be reduced, and the volume and compliance will experience an abrupt change. The structures of the modifications are shown in Fig. 8.18. Although there are no intersecting polygons inside these structures, the overhang angle condition is no longer satisfied. In order to solve this issue, a second round of optimization is further carried out. Clearly, the optimized results shown in Fig. 8.19 have no V-shapes and the inclined angles of the

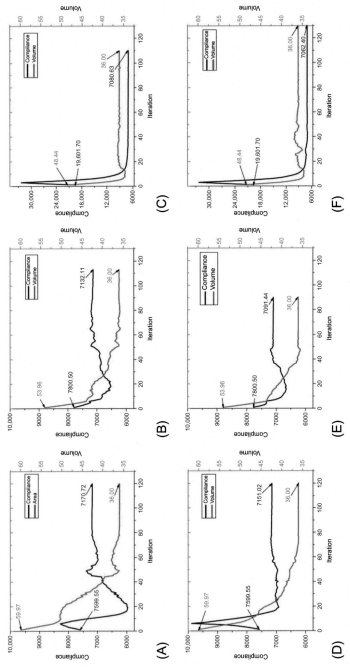

FIG. 8.16 The convergence curves of compliance and volume for freeform topology optimization: (A) case 1; (B) case 2; (C) case 3; (D) case 4; (E) case 5; and (F) case 6.

TABLE 8.2 The critical ratios for a 24-side polygon.

$\bar{\lambda}_1 = \bar{\lambda}_{24} = 0.8165$	$\bar{\lambda}_2 = \bar{\lambda}_{23} = 0.8966$
$\bar{\lambda}_3 = \bar{\lambda}_{22} = 0.9659$	$\bar{\lambda}_4 = \bar{\lambda}_{21} = 1.0353$
$\bar{\lambda}_5 = \bar{\lambda}_{20} = 1.1154$	$\bar{\lambda}_6 = \bar{\lambda}_{19} = 1.2247$
$\bar{\lambda}_7 = \bar{\lambda}_{18} = 1.4142$	$\bar{\lambda}_8 = \bar{\lambda}_{17} = 1.9319$

overhang portions are strictly restricted by the critical overhang angle. At this point, these structures can be realized by AM without adding any supports. Moreover, the final values of compliance are also found to be decreased with the increase of the total number of design variables. Compared with freeform optimized results, the compliance goes up by 12.63%, 5.52%, 5.90%, 11.73%, 5.58%, and 5.64%, respectively. Fig. 8.20 shows the convergence curves of structural compliance and volume in all cases. Notice that red lines are used to represent different round stages.

Example 8.2 A short beam with a vertical load at the middle point of the right side
Unlike the previous example, the short beam has a dimension of $L \times 5L/8$. A vertical force is applied at the middle point of the right side, as shown in Fig. 8.21. Suppose that the length $L = 8$ and the model are divided into an

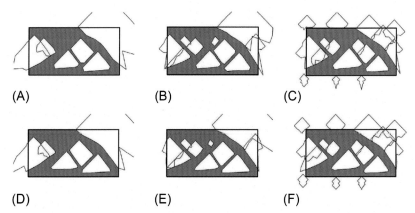

(A) (B) (C)

(D) (E) (F)

FIG. 8.17 The results of the first round topology optimization with the imposed condition $\beta \geq 45°$: (A) case 1 ($J = 7586.25$, $A = 36.00$); (B) case 2 ($J = 7377.07$, $A = 36.00$); (C) case 3 ($J = 7343.95$, $A = 36.00$); (D) case 4 ($J = 7523.41$, $A = 36.00$); (E) case 5 ($J = 7362.50$, $A = 36.00$); and (F) case 6 ($J = 7331.19$, $A = 36.00$).

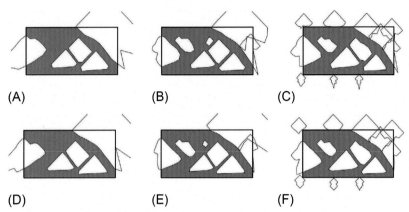

FIG. 8.18 The initial topologies of the second round topology optimization with the imposed condition $\beta \geq 45°$: (A) case 1 ($J=7145.12$, $A=38.76$); (B) case 2 ($J=7153.11$, $A=37.75$); (C) case 3 ($J=7092.67$, $A=37.92$); (D) case 4 ($J=7111.56$, $A=38.65$); (E) case 5 ($J=7519.47$, $A=35.97$); and (F) case 6 ($J=7304.32$, $A=37.78$).

80×50 FCM mesh. This example is intended to illustrate the influence of β_0 on the optimized results. Tests are made for different values $\beta_0 = 26.6°$, $45°$, $63.4°$.

First, consider the initial structure involving 17 polygons with 24 sides shown in Fig. 8.22A. The freeform optimized result and the convergence curves of compliance and volume are given in Fig. 8.22B and C, respectively. Without the restriction of the critical overhang angle, there are no V-shapes in the optimized result and this is because the V-shaped areas are not effective load paths. For each value of β_0, the boundaries requiring supports are marked in red in

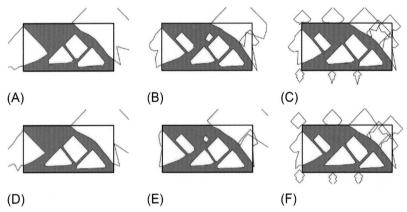

FIG. 8.19 The optimized results of the second round topology optimization with the imposed condition $\beta \geq 45°$: (A) case 1 ($J=8076.21$, $A=36.00$); (B) case 2 ($J=7526.09$, $A=36.00$); (C) case 3 ($J=7498.70$, $A=36.00$); (D) case 4 ($J=7990.82$, $A=36.00$); (E) case 5 ($J=7506.64$, $A=36.00$); and (F) case 6 ($J=7460.40$, $A=36.00$).

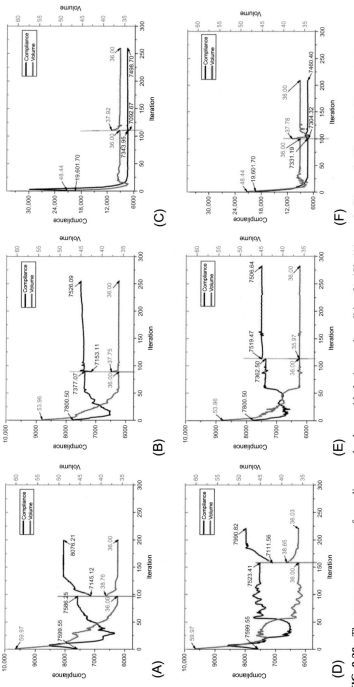

FIG. 8.20 The convergence curves of compliance and volume with the imposed condition $\beta \geq 45°$: (A) case 1; (B) case 2; (C) case 3; (D) case 4; (E) case 5; and (F) case 6.

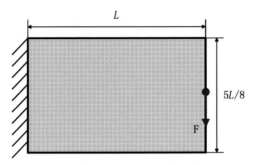

FIG. 8.21   A short beam with a vertical load at the middle point of the right side.

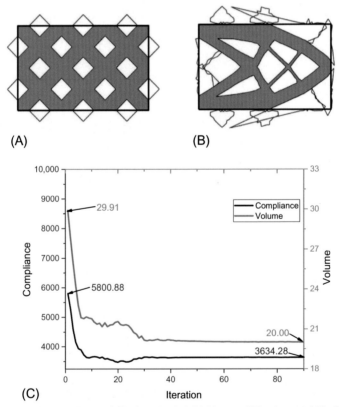

(A)   (B)

(C)

FIG. 8.22   Freeform topology optimization: (A) the initial layout of 17 polygons within the design domain ($J = 5800.88$, $A = 29.91$); (B) the optimized result ($J = 3634.28$, $A = 20.00$); and (C) the convergence curves of compliance and volume.

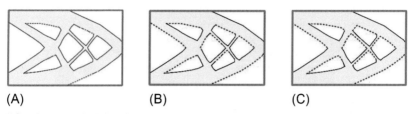

FIG. 8.23 Identified boundaries requiring supports within the freeform optimized structure for different β_0: (A) $\beta_0 = 26.6°$; (B) $\beta_0 = 45°$; and (C) $\beta_0 = 63.4°$.

TABLE 8.3 The composition of design variables for a 24-side polygon with different β_0.

COA(β_0)	Number of center coordinates	Number of ratio design variables	Number of length design variables
26.6°	2	14	10
45°	2	16	8
63.4°	2	20	4

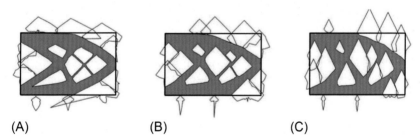

FIG. 8.24 The optimized result of the first round topology optimization: (A) $\beta_0 = 26.6°$ ($J = 3679.95$, $A = 20.00$); (B) $\beta_0 = 45°$ ($J = 3864.48$, $A = 20.00$); and (C) $\beta_0 = 63.4°$ ($J = 4539.63$, $A = 20.00$).

Fig. 8.23. In other words, the red lines indicate the violations of the critical angles when the structure is built from the bottom up. Certainly, the total length of red lines is enlarged along with the increasing value of β_0.

Table 8.3 lists the composition of design variables in a 24-side polygon with the critical overhang angles of 26.6°, 45°, and 63.4°. It is clear that the greater the values of β_0, the larger the number of ratio design variables. Fig. 8.24 shows the optimized results with the imposition of different values of β_0. It can be seen that the optimized structures are no longer symmetrical due to the imposition

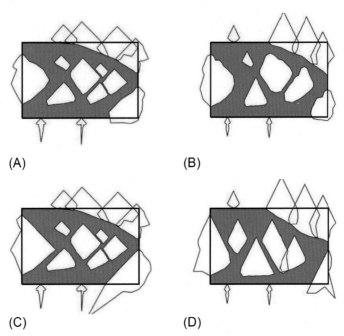

(A)   (B)

(C)   (D)

FIG. 8.25 The initial and optimized results of the second round topology optimization: (A) the initial topology with $\beta_0 = 45°$ ($J = 3673.70$, $A = 21.23$); (B) the initial topology with $\beta_0 = 63.4°$ ($J = 3944.58$, $A = 22.34$); (C) the optimized result with $\beta_0 = 45°$ ($J = 3978.24$, $A = 20.00$); and (D) the optimized result with $\beta_0 = 63.4°$ ($J = 4838.58$, $A = 20.00$).

of β_0. The least constrained case is $\beta_0 = 26.6°$ and the corresponding solution bears the closest resemblance to the solution of freeform topology optimization. Moreover, in addition to the first structure, V-shaped areas are found in the other two structures, especially in the structure of Fig. 8.24C. The corresponding modifications are shown in Fig. 8.25A and B. For $\beta_0 = 45°$, the number of polygons is reduced to 14 while there are only 12 polygons left for $\beta_0 = 26.6°$.

Realizing that the overhang angle condition is not satisfied at these newly modified polygons, a second round of topology optimization is carried out. The reoptimized results are shown in Fig. 8.25C and D. All the original V-shapes have been eliminated and the structure in Fig. 8.25C is the final solution for $\beta_0 = 45°$. In addition, Fig. 8.25D indicates that even with reoptimization, V-shape areas are still possible in the optimized result due to the intersections of other polygons. In order to get a fully self-supporting structure, the number of polygons has to be further reduced. Based on the modified structure given in Fig. 8.26A, the final solution free of V-shaped areas is illustrated in Fig. 8.26B.

It is found that the V-shaped area is more likely to appear with a larger value of β_0. Moreover, a more restrictive critical overhang angle leads to a large reduction of the number of polygons and further simplifies the definition of

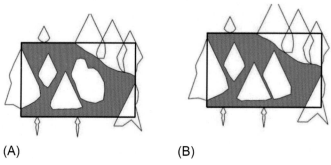

(A)   (B)

FIG. 8.26 The initial and optimized results of the third round topology optimization with $\beta_0 = 63.4°$: (A) the initial topology ($J = 4754.21$, $A = 20.23$); and (B) the optimized result ($J = 5097.31$, $A = 20.00$).

structural topology. Fig. 8.27 represents the convergence curves in all cases where the blue and green solid lines correspond to the convergence curves of compliance and volume, respectively. The red lines are used to distinguish optimization stages in different rounds. Three-round topology optimizations are experienced in the case of $\beta_0 = 63.4°$.

Supports are eliminated to achieve a self-supporting structure at the expense of reducing the structural stiffness. In detail, the freeform optimization achieves a compliance of $J = 3634.28$ while the compliance values are 3679.95, 3978.24, and 5097.31 for $\beta_0 = 26.6°$, $\beta_0 = 45°$, and $\beta_0 = 63.4°$. This corresponds to increasing values of 1.26%, 9.46%, and 40.26%. With the same constraint of β_0, the loss of stiffness is less than that of the optimized solutions that are obtained by means of the overhang projection approach (Gaynor and Guest, 2016). In the three cases of Gaynor and Guest (2016), each optimized compliance increases by 6%, 19%, and 63% over the freeform optimized result. In addition, the intermediate densities are found in some optimized results of Gaynor and Guest (2016). It is undesirable that fully dense structures are supported by intermediate density materials. But in our work, as design variables are utilized to control the shape of the polygon and the location, traditional material interpolation models are no longer needed to penalize the material properties. So, topology optimization is realized in the way of shape optimization and all optimized topologies are free of the low-density region and jagged boundary.

Example 8.3 Topology optimization of an MBB beam

In this study, an MBB beam is taken as an example to investigate the influence of build direction. As illustrated in Fig. 8.28, only half the model is considered due to the symmetry of this problem. The model is discretized by a 120×40 FCM mesh and the left side is applied to the symmetric boundary condition. Meanwhile, the displacement of the bottom right corner is not allowed in the y direction and a unit vertical load is imposed on the top left.

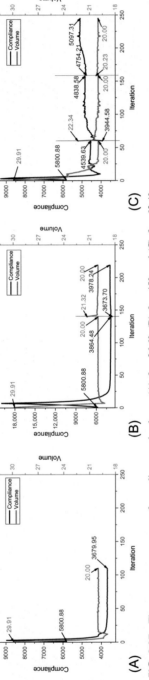

FIG. 8.27 The convergence curves of compliance and volume: (A) $\beta_0 = 26.6°$; (B) $\beta_0 = 45°$; and (C) $\beta_0 = 63.4°$.

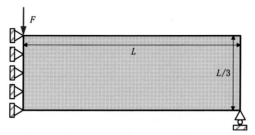

FIG. 8.28 Half of an MBB beam.

In this test, optimization is performed using four different build directions (**E, W, N, S**). As shown in Fig. 8.29A, the same initial topology structure including 16 polygons with 24 sides is studied. The freeform optimized result is illustrated in Fig. 8.29B and it cannot be printed without supports in any build direction. Fig. 8.29C represents the convergence curves of compliance and volume, respectively. Because the volume constraints are not met yet, compliance changes sharply at the beginning. The boundaries requiring support structures with different build directions are highlighted in red in Fig. 8.30. It is obvious that when the build direction is **E/W**, the total length of the boundaries violating the overhang angle condition is far less than that of the **N/S** case.

In all cases, the composition and the number of design variables in a polygon are unchanged, but the definition of direction $\overrightarrow{OV_i}$ is different. $\overrightarrow{OV_1}$ is always consistent with the build direction and the subscript of the remaining $\overrightarrow{OV_i}$ is incremented counter-clockwise.

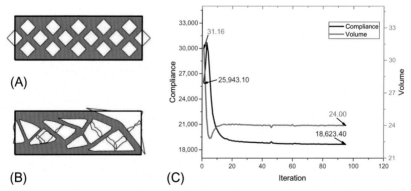

FIG. 8.29 Freeform topology optimization: (A) the initial layout of polygons within the design domain ($J = 25{,}943.10$, $A = 31.16$); (B) the optimized result ($J = 18{,}623.40$, $A = 24.00$); and (C) the convergence curves of compliance and volume.

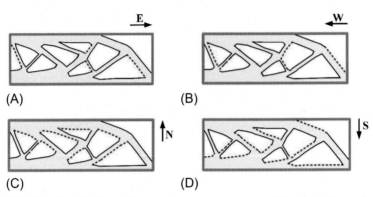

FIG. 8.30 The identified boundaries requiring support structures within the freeform optimized structure with different build directions: (A) $\mathbf{b}=\mathbf{E}$; (B) $\mathbf{b}=\mathbf{W}$; (C) $\mathbf{b}=\mathbf{N}$; and (D) $\mathbf{b}=\mathbf{S}$.

Results related to four build orientations and $\beta_0=45°$ are illustrated in Fig. 8.31. It's clear that different orientations have resulted in completely different topologies. All designs fully comply with the overhang angle condition. In particular, the structure of the case **E/W** is capable of being directly printed without additional supports. However, the results in the **N/S** case need to be modified and reoptimized due to the existence of V-shaped areas. Fig. 8.32 gives the convergence histories of compliance and volume. Compared to the freeform optimized structure, each direction **E/W/N/S** achieves an increasing compliance value by 1.16%, 2.11%, 5.22%, and 9.19%. Considering that reoptimization will further increase the compliance, the **E** direction is preferred to achieve a printable design with minimum compliance. Thus, by only using a round of topology optimization, a self-supporting structure with the least loss of stiffness is obtained.

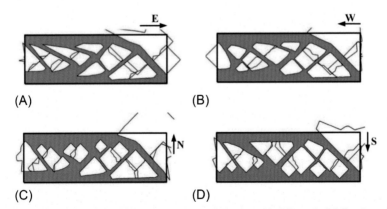

FIG. 8.31 The results of the first round topology optimization with different build directions: (A) $\mathbf{b}=\mathbf{E}$ ($J=18,840.00$, $A=23.93$); (B) $\mathbf{b}=\mathbf{W}$ ($J=19,017.20$, $A=24.00$); (C) $\mathbf{b}=\mathbf{N}$ ($J=19,595.00$, $A=24.00$); and (D) $\mathbf{b}=\mathbf{S}$ ($J=20,334.70$, $A=24.00$).

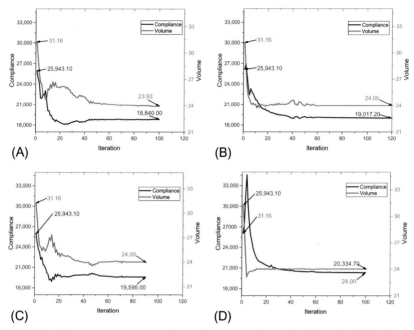

FIG. 8.32 The convergence curves of compliance and volume with different build directions: (A) **b**=**E**; (B) **b**=**W**; (C) **b**=**N**; and (D) **b**=**S**.

8.2 Topology optimization method with elimination of enclosed voids

Another critical issue in the AM process concerns the avoidance of enclosed voids inside the optimized structure for structural connectivity. The existence of enclosed voids implies that there is no way to get the unmelted powders and inner supports out of these voids after the part is completed (Diegel et al., 2010; Chua and Leong, 2014; Albakri et al., 2015; Meisel and Williams, 2015; Hu et al., 2017; Liu and Ma, 2016; Zhou and Saitou, 2017) by the AM techniques such as selective laser melting (SLM) and stereo lithography appearance (SLA).

Liu et al. (2015) and Li et al. (2016) proposed a virtual temperature method to force the connections of isolated voids for structural connectivity. In detail, a virtual heating source is introduced to each void field while solid areas are filled with thermal insulation materials. In this way, isolated voids will be controlled by constraining the maximum temperature over the structure to release the accumulation of heat energy. Clearly, this method requires an additional temperature field analysis so that the choices of temperature limit and conductivity parameters significantly influence the final design.

Moreover, the elimination of enclosed voids in a structure is also required in casting, cutting, and some other manufacturing processes. Take casting process as an example. In order to facilitate the removal of the casting mold, enclosed voids are not allowed in casting parts. As pointed out by Xia et al. (2010), a new void cannot be nucleated in the interior of a structure if the conventional level-set based method without the topological derivative is adopted. Gersborg and Andreasen (2011) implicitly involved a connectivity constraint by using a Heaviside design parameterization. These methods are always tightly coupled to other casting constraints, which will generate too conservative results. Hence, there are still a lot of spaces exploiting a better approach for structural connectivity.

In this section, the feature-driven topology optimization method is extended to take into account the structural connectivity for AM. The design procedure is as follows. Void features are modeled by level-set functions in terms of closed B-splines (CBS) or superellipses that act as basic primitives in topology optimization. A side constraint scheme is proposed to bound design variables related to center points of these features outside the contour of the design domain. This scheme implies that the avoidance of enclosed voids is realized without introducing additional nonlinear constraints and additional computing burdens. Besides, the fixed mesh technique is adopted for structural analysis and sensitivity analysis.

8.2.1 Side constraint scheme for structural connectivity

Fig. 8.33A shows an example in terms of structural connectivity without an enclosed void. In contrast, the existence of two inner voids shown in Fig. 8.33B will make it difficult to remove unmelted powders or auxiliary supports produced in the AM process. The proposed side constraint scheme is to take locations of center points of void features as design variables and then constrain their variations outside the design domain.

To make things clear, consider the rectangular domain with dimensions $L_1 \times L_2$ in Fig. 8.34. All 10 void features are constrained to impose their center locations along the borders of the design domain at least. Each feature will only have a part inside the design domain so that enclosed voids are avoided to ensure structural connectivity.

A summary is made about the side constraints to the feature centers in Table 8.4. It is noteworthy that there are two options for the void features with centers located at the corners. The same method can be extended to 3D problems. Void features are initially distributed on the boundary surfaces of the design domain and the variation bounds of design variables are determined correspondingly.

From the above presentation, we can see that structural connectivity can be effectively achieved without introducing any additional constraint in topology optimization. In this way, only proper bound values are imposed for design variables related to the void centers within the framework of the feature-driven method.

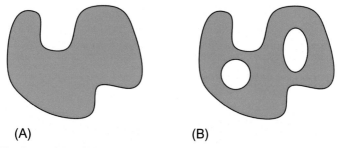

FIG. 8.33 Connectivity of 2D structures: (A) structure without enclosed void; and (B) structure with two enclosed voids.

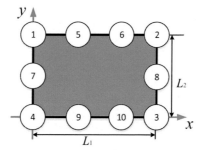

FIG. 8.34 Initial structure with 10 void features.

TABLE 8.4 Specific requirements for void features.

Void features	Requirements	Void features	Requirements
1	$x_0 \leq 0$ or $y_0 \geq L_2$	6	$y_0 \geq L_2$
2	$x_0 \geq L_1$ or $y_0 \geq L_2$	7	$x_0 \leq 0$
3	$x_0 \geq L_1$ or $y_0 \leq 0$	8	$x_0 \geq L_1$
4	$x_0 \leq 0$ or $y_0 \leq 0$	9	$y_0 \leq 0$
5	$y_0 \geq L_2$	10	$y_0 \leq 0$

8.2.2 Numerical examples

In this section, several numerical examples are provided to validate the effectiveness of the proposed method. The Young's modulus of solid material and Possion's ratio are $E_0 = 1$ and $\nu = 0.3$, respectively. The geometric data and loads are all dimensionless in the following examples.

Example 8.4 Topology optimization of a short beam

Fig. 8.35 depicts a short beam. It is completely fixed along the left side and a vertical force is applied at the middle point of the right side. The model is discretized into an 80×40 quadrilateral mesh. The volume fraction of 50% is used as the upper bound of the volume constraint.

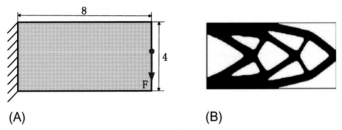

FIG. 8.35 A short beam: (A) design domain; and (B) the freely optimized result (Luo et al., 2009).

First, CBS curves are considered as void features to realize topological changes. In order to study the influence of connectivity constraint upon the optimized topologies, both freeform topology optimization and topology optimization considering structural connectivity are carried out. Figs. 8.36A and 8.37A show the corresponding initial design models composed of 17 voids, each of which has 24 control radii. In total, 442 design variables exist. Fig. 8.36B gives the freely optimized result, which shares the same topology as in Fig. 8.35B. Six inner holes violate the connectivity constraint. It is also observed that some void features move outside the design domain and become useless. Fig. 8.36C depicts the convergence curves of compliance and volume.

When the void features in Fig. 8.37A are constrained to limit their locations of center points, each feature will have a portion that cannot enter the interior of the design domain. Hence, all voids are guaranteed to communicate with the outside of the design domain. The optimized result with connectivity is illustrated in Fig. 8.37B. The solution is a simply connected structure to favor the evacuation of the unmelted powders or liquids and removing support materials during the AM process. To have a clear idea about the topological changes, some intermediate iterations are given in Table 8.5. It is observed that structural optimization considering connectivity is actually a process of changing the boundary of the design domain. The convergence histories of compliance and volume are shown in Fig. 8.37C. In comparison, the freeform optimization achieves a compliance of 6011.72 while the compliance of the structure with connectivity has an increasing value of 19.78%.

Now, the same number of superellipses is used to drive topology optimization. Each superellipse has five design variables related to the position, orientation, semilength, and semiwidth. There are a total of 85 design variables, which is only one-fifth of the CBS features. Fig. 8.38A represents the initial

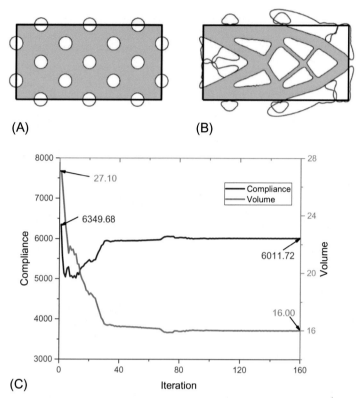

FIG. 8.36 Freeform topology optimization with CBS void features: (A) the initial layout ($J = 6349.68$, $V = 27.10$); (B) the optimized result ($J = 6011.72$, $V = 16.00$); and (C) the convergence curves of compliance and volume.

configuration without considering structural connectivity. Compared with the CBS feature, the superellipse tends to achieve topological change through intersection because of its insufficient deformation ability. This leads to a relatively simple topology with three inner holes in Fig. 8.38B. Fig. 8.38C represents the convergence curves of compliance and volume. Usually, this deficiency is compensated by increasing the number of superellipses. When the initial structure consists of 34 superellipses distributed in a crossing way over the design domain, the corresponding optimized structure shares the topology similar to the result in Fig. 8.36B, as illustrated in Fig. 8.39.

For topology optimization considering structural connectivity, superellipses are initially distributed along the boundary of the design domain to facilitate the constraints on center points. Figs. 8.40A and 8.41A give two initial structures with different numbers of superellipses. However, Figs. 8.40B and 8.41B illustrate that both the optimized structures are similar to the CBS-based result without an enclosed void shown in Fig. 8.37B. The values of structural compliance

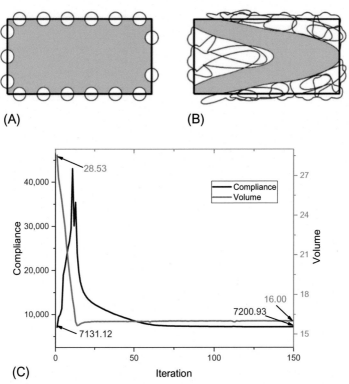

FIG. 8.37 Topology optimization considering structural connectivity with CBS void features: (A) the initial layout ($J=7131.12$, $V=28.53$); (B) the optimized result ($J=7200.93, V=16.00$); and (C) the convergence curves of compliance and volume.

correspond to 7880.14 and 7446.87. The convergence histories are shown in Figs. 8.40C and 8.41C, respectively.

Example 8.5 Topology optimization of a simply supported hexahedron
A 3D hexahedron studied in Ref. (Li et al., 2016) is considered in Fig. 8.42. It has a dimension of $40 \times 40 \times 20$ with four corners fixed at the bottom face in all three directions. A vertical force is applied at the center of the bottom face. Suppose that a bottom layer of $40 \times 40 \times 2$ is a nondesignable solid domain. Here, the structure is discretized with $50 \times 50 \times 25$ eight-node hexahedra elements. The volume fraction is limited to 30%, equally a volume of 9600.

Three cases are considered in Table 8.6 and initial layouts of CBS voids are shown in Fig. 8.43 to study the influences upon the optimized topology. Each void feature is represented by the CBS defined by 120 control radii. The initial structures will be used to perform both topology optimization without and with the consideration of structural connectivity, respectively. For the former, features are allowed to move freely rather than being restricted outside the design

TABLE 8.5 Evolutions of structural topology with connectivity.

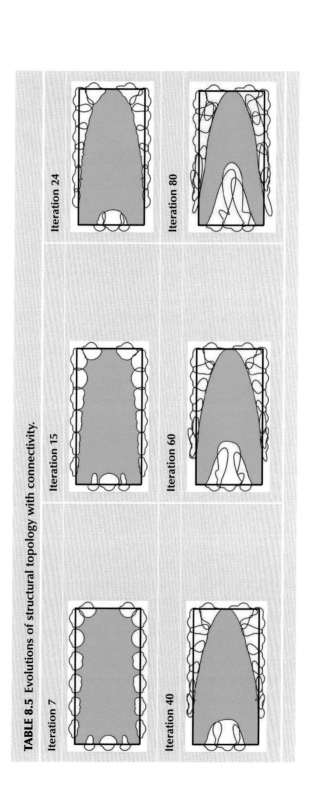

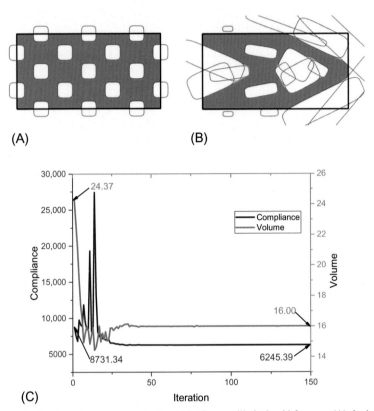

FIG. 8.38 Freeform topology optimization with 17 superelliptical void features: (A) the initial layout ($J = 8731.34$, $V = 24.37$); (B) the optimized result ($J = 6245.39$, $V = 16.00$); and (C) the convergence curves of compliance and volume.

domain. Fig. 8.44A–C gives the results of freeform optimization in three cases. Correspondingly, Fig. 8.44D–F provides the half models for the better observation of interior topologies. It can clearly be seen that even with different numbers of void features, the similar topology with one enclosed void is produced. Besides, compliance decreases as the number of features increase, but the amount of reduction is small.

Now, side constraints are imposed for design variables related to the center coordinates of featured CBS. In detail, the z-coordinates of center points are all bounded by $z_0 \geq 20$ for the features centered on the top surface. Similarly, other features are constrained for the x- or y-coordinates of center points. The optimized results and the corresponding half models are shown in Fig. 8.45. In three cases, the enclosed holes disappear. The compliance values are 239.71, 238.76, and 237.06 with the increasing values of 2.56%, 2.69%, and 2.60%.

In the results of freeform topology optimization, distributions of the center points of CBS features are depicted in Fig. 8.46. Black squares representing the

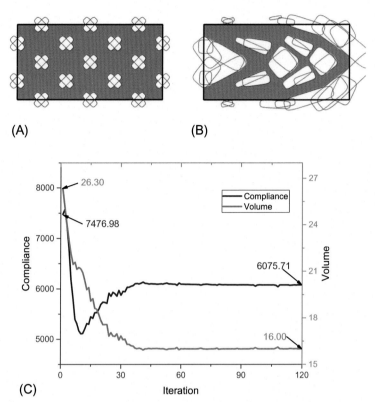

FIG. 8.39 Freeform topology optimization with 34 superelliptical void features: (A) the initial layout ($J = 7476.98$, $V = 26.30$); (B) the optimized result ($J = 6075.71, V = 16.00$); and (C) the convergence curves of compliance and volume.

center points inside the domain have a large number while red solid circles representing the center points located outside the design domain are few. By contrast, all center points are successfully restricted outside the design domain owing to the imposed side constraints for the connectivity, as illustrated in Fig. 8.47. For two kinds of topology optimization, the convergence curves of compliance and volume are shown in Figs. 8.48 and 8.49, respectively. At the beginning of the iterations, the compliance changes sharply because the volume constraint is violated.

Consider now the superellipse as an alternative feature. The initial structure is shown in Fig. 8.50 and the dimension of each void feature is $3 \times 2 \times 2$. There are 65 superellipses perpendicular to the surfaces of the design domain. The total number of design variables is thus $65 \times 9 = 585$. Fig. 8.51A and B shows the optimized result and the corresponding half model, respectively. In

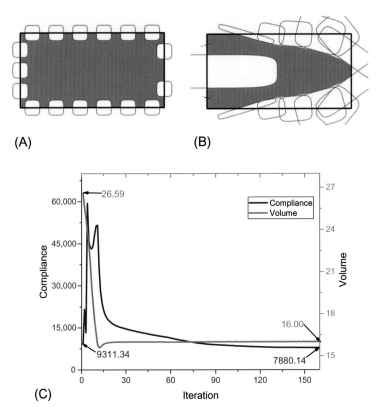

FIG. 8.40   Topology optimization considering structural connectivity with 17 superelliptical void features: (A) the initial layout ($J=9311.34$, $V=26.59$); (B) the optimized result ($J=7880.14$, $V=16.00$); and (C) the convergence curves of compliance and volume.

comparison with the CBS-based results, Fig. 8.51A produces a similar topology with one enclosed void.

Fig. 8.52A represents the optimized result with the imposition of side constraints. From the half model shown in Fig. 8.52B, it can clearly be seen that no inner void exists. The compliance is 239.40, which is 2.96% higher than the freely optimized result ($J=232.51$). Fig. 8.53A and B depicts the distributions of the center points of superellipses in both optimized results. The convergence histories of compliance and volume are compared in Fig. 8.54A and B.

Example 8.6 Topology optimization of a torsion beam

Another 3D example of a torsion beam shown in Fig. 8.55 is studied. The structure is completely fixed along the left side and four loads are imposed on the four vertices of the right plane with an inclined angle of 45°. Suppose two cuboid zones with a dimension of $20 \times 20 \times 2$ at both ends of the structure are chosen as nondesign solid domains. The beam is discretized into hexahedral

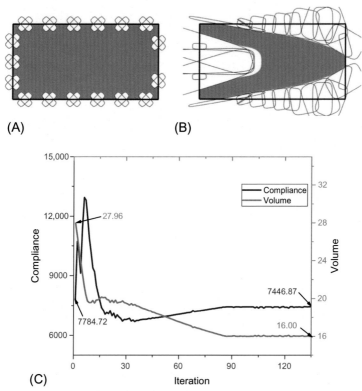

(A)   (B)

(C)

FIG. 8.41 Topology optimization considering structural connectivity with 34 superelliptical void features: (A) the initial layout ($J = 7784.72$, $V = 27.96$); (B) the optimized result ($J = 7446.87$, $V = 16.00$); and (C) the convergence curves of compliance and volume.

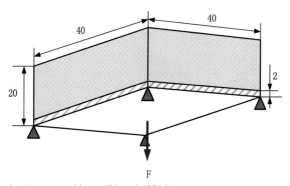

FIG. 8.42 A simply supported beam (Li et al., 2016).

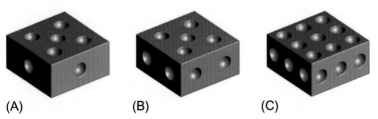

FIG. 8.43 The initial structure with CBS void features: (A) nine void features ($J=218.47$, $V=30{,}782.70$); (B) 13 void features ($J=219.07, V=30{,}262.40$); and (C) 21 void features ($J=220.28, V=29{,}158.90$).

TABLE 8.6 Three design cases of initial structure.

Items	The number of void features on the top surface	The number of void features on the four side surfaces	The total number of void features
Case 1	5	4	9
Case 2	5	8	13
Case 3	9	12	21

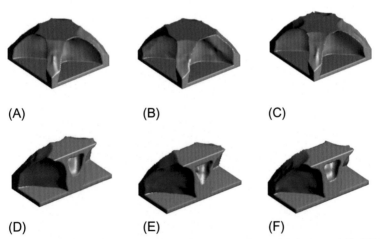

FIG. 8.44 Freeform topology optimization: (A) the optimized result in case 1 ($J=233.72$, $V=9599.00$); (B) the optimized result in case 2 ($J=232.50, V=9599.72$); (C) the optimized result in case 3 ($J=231.05, V=9600.00$); (D) half the optimized result in case 1; (E) half the optimized result in case 2; and (F) half the optimized result in case 3.

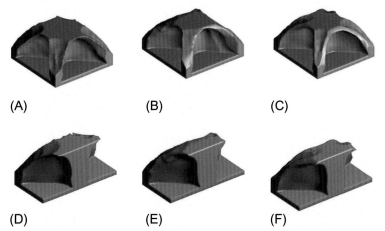

FIG. 8.45 Topology optimization considering structural connectivity: (A) the optimized result in case 1 ($J = 239.71$, $V = 9598.37$); (B) the optimized result in case 2 ($J = 238.78$, $V = 9598.71$); (C) the optimized result in case 3 ($J = 237.06$, $V = 9599.20$); (D) half the optimized result in case 1; (E) half the optimized result in case 2; and (F) half the optimized result in case 3.

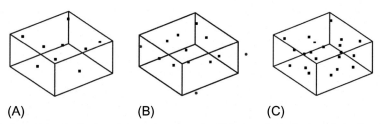

FIG. 8.46 The distribution of the center points of void features in the freely optimized results: (A) case 1; (B) case 2; and (C) case 3.

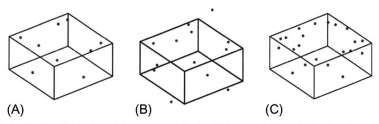

FIG. 8.47 The distribution of the center points of void features in the optimized results considering structural connectivity: (A) case 1; (B) case 2; and (C) case 3.

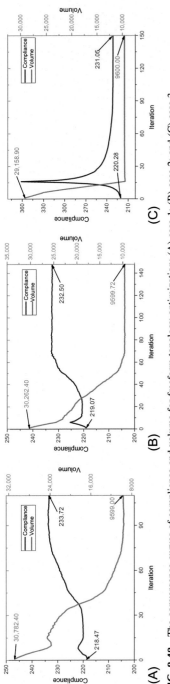

FIG. 8.48 The convergence curves of compliance and volume for freeform topology optimization: (A) case 1; (B) case 2; and (C) case 3.

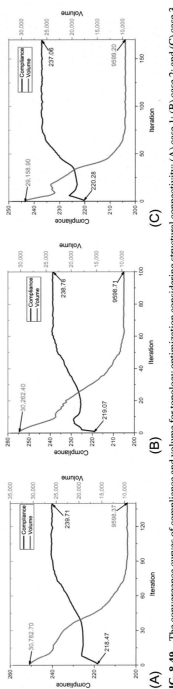

FIG. 8.49 The convergence curves of compliance and volume for topology optimization considering structural connectivity: (A) case 1; (B) case 2; and (C) case 3.

FIG. 8.50   The initial structure with void features of superellipse ($J = 226.69$, $V = 29,185.60$).

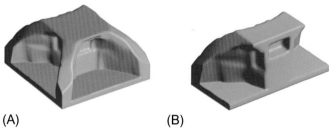

(A)   (B)

FIG. 8.51   Freeform topology optimization: (A) the optimized result ($J = 232.51$, $V = 9600.00$); and (B) half of the optimized result.

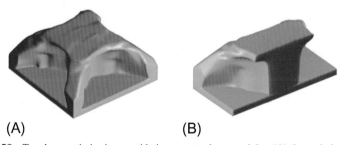

(A)   (B)

FIG. 8.52   Topology optimization considering structural connectivity: (A) the optimized result ($J = 239.40$, $V = 9599.95$); and (B) half of the optimized result.

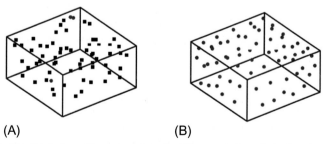

(A)   (B)

FIG. 8.53   The distribution of the center points of void features in the optimized results: (A) freely optimized result; and (B) optimized result of connectivity.

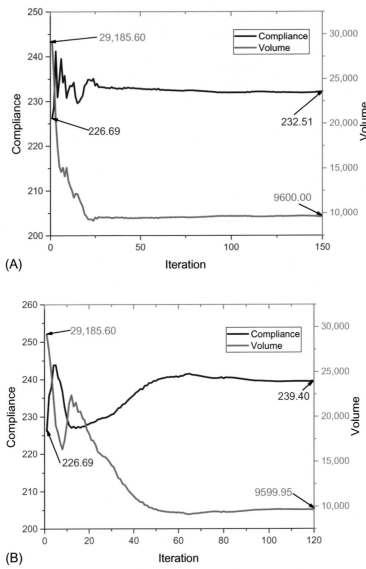

FIG. 8.54 The convergence curves of compliance and volume: (A) freeform topology optimization; and (B) topology optimization considering structural connectivity.

elements of size $0.8 \times 0.8 \times 0.8$. The upper bound of the volume is set to be 7200 with a volume fraction of 30%.

As shown in Fig. 8.56, there are 44 CBS void features with 60 controlling radii in the initial structure and all features are centered on the surfaces. The total number of design variables is $44 \times (60 + 3) = 2772$. To highlight the effect

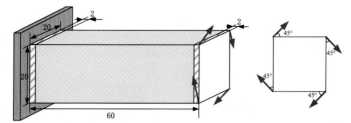

FIG. 8.55   A torsion beam (Liu et al., 2015).

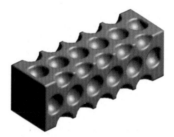

FIG. 8.56   The initial structure with CBS void features.

of semibandwidth Δ for the narrow-band sensitivity analysis scheme in Section 5.3.1, Fig. 8.57A and B gives the material distributions of the initial structure when Δ is set to 0.1 and 0.4, respectively. The latter has about one layer of intermediate elements around each hole, but only a few scattered intermediate elements exist in the former. The corresponding freely optimized results are illustrated in Fig. 8.58A and B. The half models in Fig. 8.58C and D indicates that a large enclosed void exists. Besides, compared with Fig. 8.58B, there are many pits on the surfaces of the structure in Fig. 8.58A.

In Fig. 8.56, the center points of all void features are restricted outside the design domain. The optimized results and the half models are shown in Fig. 8.59. Although structure configurations are no longer the same with different values of Δ, holes are always generated on the outer surface to realize structural connectivity. Similarly, the optimized result tends to have rough surfaces with $\Delta = 0.1$. Compared with their respective freely optimized results, the compliance of the structure with connectivity goes up by 72.56% and 13.07%. Figs. 8.60 and 8.61 gives the convergence histories of two kinds of topology optimization. The curves with $\Delta = 0.1$ always fluctuate greatly due to the sensitivity inaccuracy caused by too few intermediate elements. Therefore, a too small value of Δ causes a deterioration of structure smoothness and large fluctuations of the convergence curves.

With a better choice ($\Delta = 0.4$), a topology optimization with void features of the superellipse is also performed. The initial distribution of superellipses is demonstrated in Fig. 8.62. There are 120 superellipses centered on the surfaces

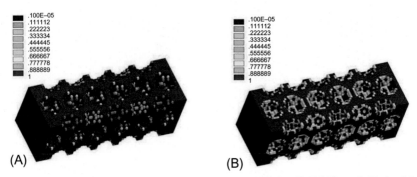

FIG. 8.57 Material distributions: (A) $\Delta=0.1$ ($J=10,376.80$, $V=19,435.10$); and (B) $\Delta=0.4$ ($J=10,217.00$, $V=19,417.60$).

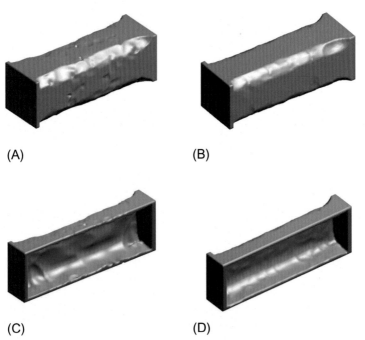

FIG. 8.58 Freeform topology optimization: (A) the optimized result with $\Delta=0.1$ ($J=10,844.80$, $V=7186.29$); (B) the optimized result with $\Delta=0.4$ ($J=10,614.00$, $V=7199.83$); (C) half of the optimized structure with $\Delta=0.1$; and (D) half of the optimized structure with $\Delta=0.4$.

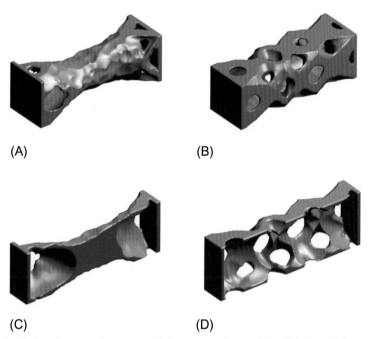

(A)   (B)

(C)   (D)

FIG. 8.59 Topology optimization considering structural connectivity: (A) the optimized result with $\Delta = 0.1$ ($J = 18{,}714.10$, $V = 7199.06$); (B) the optimized result with $\Delta = 0.4$ ($J = 12{,}001.20$, $V = 7198.67$); (C) half of the optimized structure with $\Delta = 0.1$; and (D) half of the optimized structure with $\Delta = 0.4$.

of the design domain. The total number of design variables is $120 \times 9 = 1080$, which is about one-third of the CBS-based topology optimization. Optimized topology is obtained after the movements, rotations, and deformations of super-ellipses, and this is shown in Fig. 8.63A. The corresponding half-model is illustrated in Fig. 8.63B. The big cavity is enclosed, which inevitably leads to the accumulation of a large amount of powders or liquids.

Fig. 8.64A represents the optimized structure considering connectivity and its half-model is provided in Fig. 8.64B for a better observation. Clearly, it is the desired topology that has many accesses for the support structures or unmelted powers to be moved out. The final compliance corresponds to $J = 12{,}284.76$ and it is 12.67% higher than that of the unconstrained optimal solution ($J = 10{,}902.80$). This is the price that should be paid for structural connectivity. The convergence histories of compliance and volume are shown in Fig. 8.65 with the satisfaction of the prescribed volume constraint.

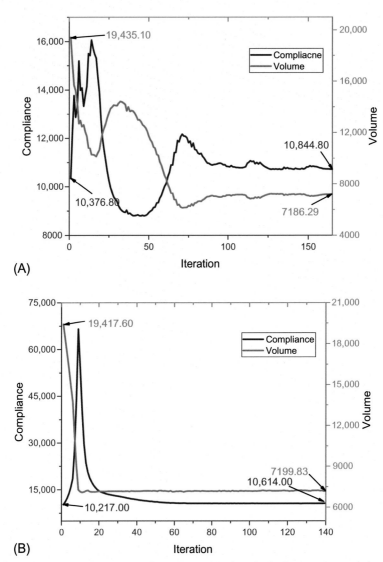

(A)

(B)

FIG. 8.60   The convergence curves of compliance and volume for freeform topology optimization: (A) $\Delta = 0.1$ and (B) $\Delta = 0.4$.

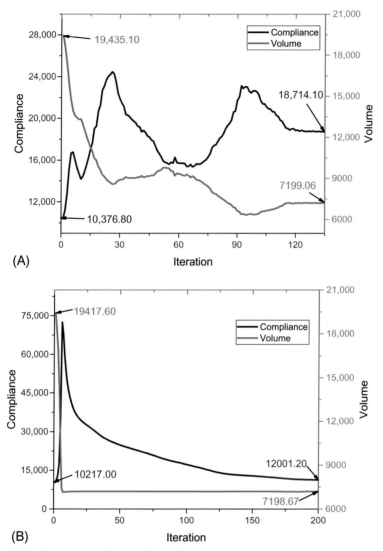

FIG. 8.61   The convergence curves of compliance and volume for topology optimization considering structural connectivity: (A) $\Delta = 0.1$; and (B) $\Delta = 0.4$.

FIG. 8.62   The initial structure with void features of superellipses ($J = 13,021.15$, $V = 15,400.96$).

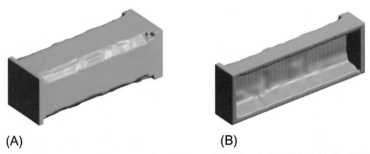

(A)   (B)

FIG. 8.63   Freeform topology optimization: (A) the optimized result ($J = 10,902.80$, $V = 7199.69$); and (B) half of the optimized structure.

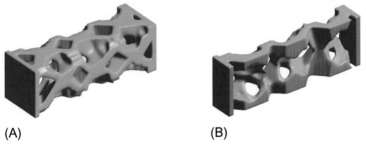

(A)   (B)

FIG. 8.64   Topology optimization considering structural connectivity: (A) the optimized result ($J = 12,284.76$, $V = 7198.39$); and (B) half of the optimized structure.

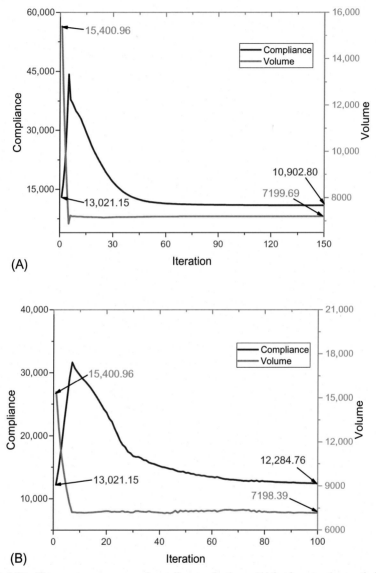

FIG. 8.65 The convergence curves of compliance and volume: (A) freeform topology optimization; and (B) topology optimization considering structural connectivity.

References

Albakri, M., Sturm, L., Williams, C.B., Tarazaga, P., 2015. Non-destructive evaluation of additively manufactured parts via impedance-based monitoring. In: International Solid Freeform Fabrication Symposium, Austin, TX, pp. 1475–1490.

Barnett, E., Gosselin, C., 2015. Weak support material techniques for alternative additive manufacturing materials. Addit. Manuf. 8, 95–104.

Brackett, D., Ashcroft, I., Hague, R., 2011, August. Topology optimization for additive manufacturing. In: Proceedings of the Solid Freeform Fabrication Symposium, Austin, TX. vol. 1. pp. 348–362.

Calignano, F., 2014. Design optimization of supports for overhanging structures in aluminum and titanium alloys by selective laser melting. Mater. Des. 64, 203–213.

Chua, C.K., Leong, K.F., 2014. 3D Pinting and Additive Manufacturing: Principles and Applications (With Companion Media Pack) of Rapid Prototyping, fourth ed. World Scientific Publishing Co. Pte. Ltd., Singapore.

Clausen, A., 2016. Topology Optimization for Additive Manufacturing. Technical University of Denmark. DCAMM Special Report, No. 214.

Das, P., Chandran, R., Samant, R., Anand, S., 2015. Optimum part build orientation in additive manufacturing for minimizing part errors and support structures. Procedia Manuf. 1, 343–354.

Diegel, O., Singamneni, S., Reay, S., Withell, A., 2010. Tools for sustainable product design: additive manufacturing. J. Sustain. Dev. 3 (3), 68–75.

Dumas, J., Hergel, J., Lefebvre, S., 2014. Bridging the gap: automated steady scaffoldings for 3D printing. ACM Trans. Graph. 33 (4), 1–10.

Frazier, W.E., 2014. Metal additive manufacturing: a review. J. Mater. Eng. Perform 23 (6), 1917–1928.

Gao, W., Zhang, Y., Ramanujan, D., Ramani, K., Chen, Y., Williams, C.B., Wang, C., Shin, Y., Zhang, S., Zavattieri, P.D., 2015. The status, challenges, and future of additive manufacturing in engineering. Comput. Aided Des. 69, 65–89.

Gaynor, A.T., Guest, J.K., 2016. Topology optimization considering overhang constraints: eliminating sacrificial support material in additive manufacturing through design. Struct. Multidiscip. Optim. 54 (5), 1157–1172.

Gaynor, A.T., Meisel, N.A., Williams, C.B., Guest, J.K., 2014. Topology optimization for additive manufacturing: considering maximum overhang constraint. In: 15th AIAA/ISSMO Multidisciplinary Analysis and Optimization Conference, p. 2036.

Gersborg, A.R., Andreasen, C.S., 2011. An explicit parameterization for casting constraints in gradient driven topology optimization. Struct. Multidiscip. Optim. 44 (6), 875–881.

Guo, X., Zhou, J., Zhang, W., Du, Z., Liu, C., Liu, Y., 2017. Self-supporting structure design in additive manufacturing through explicit topology optimization. Comput. Methods Appl. Mech. Eng. 323, 27–63.

Horn, T.J., Harrysson, O.L., 2012. Overview of current additive manufacturing technologies and selected applications. Sci. Prog 95 (3), 255–282.

Hu, K., Jin, S., Wang, C.C., 2015. Support slimming for single material based additive manufacturing. Comput. Aided Des. 65, 1–10.

Hu, R., Chen, W., Li, Q., Liu, S., Zhou, P., Dong, Z., Kang, R., 2017. Design optimization method for additive manufacturing of the primary mirror of a large-aperture space telescope. J. Aerosp. Eng. 30 (3), 04016093.

Hussein, A., Hao, L., Yan, C., Everson, R., Young, P., 2013. Advanced lattice support structures for metal additive manufacturing. J. Mater. Process. Technol. 213 (7), 1019–1026.

Kranz, J., Herzog, D., Emmelmann, C., 2015. Design guidelines for laser additive manufacturing of lightweight structures in TiAl6V4. J. Laser Appl. 27 (S1), S14001.

Langelaar, M., 2016. Topology optimization of 3D self-supporting structures for additive manufacturing. Addit. Manuf. 12, 60–70.

Langelaar, M., 2017. An additive manufacturing filter for topology optimization of print-ready designs. Struct. Multidiscip. Optim. 55 (3), 871–883.

Leary, M., Merli, L., Torti, F., Mazur, M., Brandt, M., 2014. Optimal topology for additive manufacture: a method for enabling additive manufacture of support-free optimal structures. Mater. Des. 63, 678–690.

Li, Q., Chen, W., Liu, S., Tong, L., 2016. Structural topology optimization considering connectivity constraint. Struct. Multidiscip. Optim. 54 (4), 971–984.

Liu, J., Ma, Y., 2016. A survey of manufacturing oriented topology optimization methods. Adv. Eng. Softw. 100, 161–175.

Liu, S., Li, Q., Chen, W., Tong, L., Cheng, G., 2015. An identification method for enclosed voids restriction in manufacturability design for additive manufacturing structures. Front. Mech. Eng. 10 (2), 126–137.

Luo, Z., Tong, L., Kang, Z., 2009. A level set method for structural shape and topology optimization using radial basis functions. Comput. Struct. 87 (7–8), 425–434.

Meisel, N., Williams, C., 2015. An investigation of key design for additive manufacturing constraints in multimaterial three-dimensional printing. J. Mech. Des. 137 (11), 111406.

Mertens, R., Clijsters, S., Kempen, K., Kruth, J.P, 2014. Optimization of scan strategies in selective laser melting of aluminum parts with downfacing areas. J. Manuf. Sci. Eng. 136 (6), 061012.

Mirzendehdel, A.M., Suresh, K., 2016. Support structure constrained topology optimization for additive manufacturing. Comput. Aided Des. 81, 1–13.

Morgan, H.D., Cherry, J.A., Jonnalagadda, S., Ewing, D., Sienz, J., 2016. Part orientation optimisation for the additive layer manufacture of metal components. Int. J. Adv. Manuf. Technol. 86 (5–8), 1679–1687.

Qattawi, A., Ablat, M.A., 2017. Design consideration for additive manufacturing: fused deposition modelling. Open J. Appl. Sci. 7 (6), 291–318.

Strano, G., Hao, L., Everson, R.M., Evans, K.E., 2013. A new approach to the design and optimisation of support structures in additive manufacturing. Int. J. Adv. Manuf. Technol. 66 (9–12), 1247–1254.

Vanek, J., Galicia, J.A.G., Benes, B., 2014. Clever support: efficient support structure generation for digital fabrication. Comput. Graph. Forum 33 (5), 117–125.

Wong, K.V., Hernandez, A., 2012. A review of additive manufacturing. ISRN Mech. Eng. 2012, 208760.

Xia, Q., Shi, T., Wang, M.Y., Liu, S., 2010. A level set based method for the optimization of cast part. Struct. Multidiscip. Optim. 41 (5), 735–747.

Zegard, T., Paulino, G.H., 2016. Bridging topology optimization and additive manufacturing. Struct. Multidiscip. Optim. 53 (1), 175–192.

Zhou, Y., Saitou, K., 2017. Gradient-based multi-component topology optimization for additive manufacturing (MTO-A). In: ASME 2017 International Design Engineering Technical Conferences and Computers and Information in Engineering Conference. American Society of Mechanical Engineers Digital Collection.

Index

Note: Page numbers followed by *f* indicate figures and *t* indicate tables.

Printed in the United States
By Bookmasters